AF598072

Reconfigurable Networks-on-Chip

Sao-Jie Chen · Ying-Cherng Lan
Wen-Chung Tsai · Yu-Hen Hu

Reconfigurable Networks-on-Chip

Springer

Sao-Jie Chen
National Taiwan University
Taipei
Taiwan R.O.C
E-mail: csj@cc.ee.ntu.edu.tw

Ying-Cherng Lan
National Taiwan University
Taipei
Taiwan R.O.C

Wen-Chung Tsai
National Taiwan University
Taipei
Taiwan R.O.C

Yu-Hen Hu
University of Wisconsin-Madison
Madison, WI, USA

ISBN 978-1-4419-9340-3 e-ISBN 978-1-4419-9341-0
DOI 10.1007/978-1-4419-9341-0
Springer New York Dordrecht Heidelberg London

Library of Congress Control Number: 2011943345

Cover design: eStudio Calamar S.L.

Printed on acid-free paper

Springer is part of Springer Science+Business Media (www.springer.com)

Foreword

The search for efficient communication schemes in multiprocessor systems on chips (MPSoCs) has permeated the R&D efforts of the last decade. Among various schemes, Networks on Chips (NoCs) have emerged as the preferred paradigm for on-chip communication in both the case of homogeneous multiprocessing computing systems and in the case of heterogeneous systems for embedded applications. This research and development of NoC technology has been extremely fast and it is showing its results: seven out of the top ten semiconductor manufacturers have announced products that use NoCs as a structured high-level interconnect by 2011.

This book proposes an excellent review of research in design and applications of NoCs. In particular it stresses three axes of research that are the key to the effective use of NoCs. First and foremost, the establishment of high-performance on-chip communication structures, to match the data rate produced and consumed by increasingly faster computational cores. Second, the important study of fault tolerance in networks on chips and the means to increase MPSoC reliability through NoCs. Last but not least, the analysis of energy consumption in NoCs, and the study of the trade-off energy vs. performance, which is motivated by the wider and wider presence of complex chips in portable devices.

Overall this book shows important advances over the state of the art that will affect future system design as well as R&D in tools and methods for NoC design. It represents an important reference point for both designers and electronic design automation researchers and developers.

EPFL, Lausanne, 2011 Giovanni De Micheli

Preface

Networks-on-Chip (NoC) is an emerging on-chip interconnection centric platform that leverages modern high speed communication infrastructure to mitigate the ever increasing on-chip communication challenges of modern many-core System-on-Chip (SoC) designs. Continuing shrinkage of feature dimensions of nano-scale semiconductor devices has raised grave concerns of the reliability, signal integrity, and quality of services (QoS) of traditional bus-based on-chip interconnect infrastructure. NoC represents a major paradigm shift to address these concerns by incorporating state-of-the-art high-speed data network components (such as routers and switches) and packet-based routing protocols into a novel on-chip network infrastructure. The aims of NoC developments are to provide a reliable on-chip communication platform to facilitate scalable giga-scale SoC design.

Over the past decades, numerous ground-breaking NoC-related platform development, component innovation, algorithm and protocol refinement have been reported in the form of journal and conference publications, as well as a couple of edited monographs. This book represents a succinct summary of NoC research outcome of an international collaboration team over the past few years. The emphasis of this book is on the QoS aspects of NoC development. Specifically, a case study of a bidirectional NoC architecture is discussed and several QoS issues, including performance, fault tolerance, and energy-aware computing, are carefully elaborated. The authors feel that this book would be a good complement of existing NoC literatures and should provide a practical implementation perspective of NoC which has not been fully explored in present NoC publications.

This book is organized into three integral parts. In Part I, a brief introduction of the key notions of NoC is provided. In addition to motivations and rationales, a comprehensive of preliminary background materials are provided to give a quick overview of relevant subject materials. In Part II, performance enhancing NoC network design methodologies, including routing, fault tolerance, energy awareness, and task scheduling are discussed. In Part III, a case study of a bidirectional link based NoC platform architecture is described. Its design rationale, performance enhancement characteristics, fault tolerance, and energy awareness features are also carefully analyzed.

This monograph is the outcome of a collaborative work written for researchers who are interested in learning the fundamental theories, architectures, and algorithms of NoC, as well as those who want to acquire the state-of-the-art NoC development. The authors would like to sincerely thank the foreword writer, Professor and Director Giovanni De Micheli at the In-stitute of Electrical Engineering and the Integrated Systems Center at the École Polytech-nique Fédérale de Lausanne, Switzerland.

Taipei, Taiwan — Sao-Jie Chen
Taipei, Taiwan — Ying-Cherng Lan
Taipei, Taiwan — Wen-Chung Tsai
Madison, Wisconsin — Yu-Hen Hu

Contents

Part I
Introduction to Network-on-Chip

Chapter 1
Communication Centric Design

As the density of VLSI design increases, the complexity of each component in a system raises rapidly. To accommodate the increasing transistor density, higher operating frequencies, and shorter time-to-market pressure, multi-processor System-on-Chip (MP-SoC) architectures, which use bus structures for on-chip communication and integrate complex heterogeneous functional elements on a single die, are more and more required in today's semiconductor industry. However, today's SoC designers face a new challenge in the design of the on-chip interconnects beyond the evolution of an increasing number of processing elements. Traditional bus-based communication schemes, which lack of scalability and predictability, are not capable to keep up with the increasing requirements of future SoCs in terms of performance, power, timing closure, scalability, and so on. To meet the design productivity and signal integrity challenges of next-generation system designs, a structured and scalable interconnection architecture, Network-on-Chip (NoC), has been proposed recently to mitigate the complex on-chip communication problem.

1.1 Communications-Centric Design Concept

An application can be represented as a set of computational units that require a set of communication blocks to pass information between the units. To distinguish the performance impact of these two major components, computation time is dominated by gate delay whereas communication time is dominated by wire delay. When the amount of computational units is few, the communication blocks can be done on an ad-hoc basis. However, with the shrinking size of transistors in recent years, gate delay is ever decreasing with respect to wire delay. Thus, we need a structured and scalable on-chip communication architecture to fit the increasingly complex applications on a single chip. This translates to the design of on-chip

S.-J. Chen et al., *Reconfigurable Networks-on-Chip*,
DOI: 10.1007/978-1-4419-9341-0_1,

communications architecture as being more and more important, and promotes the design concept from computation-centric design to communication-centric design.

1.1.1 Multi-Processor System-on-Chip

System-on-Chip (SoC) is an architectural concept developed in the last few decades, in which a processor or few processors along with memory and an associated set of peripherals connected by busses are all implemented on a single chip. According to the Moore's law, the trend toward many-core processing chips is now a well established one. Power-efficient processors combined with hardware accelerators are the preferred choice for most designers to deliver the best tradeoff between performance and power consumption, since computational power increases exponentially according to the calculation of dynamic power dissipation [1]. Therefore, this trend dictates spreading the application tasks into multiple processing elements where (1) each processing element can be individually turned on or off, thereby saving power; (2) each processing element can run at its own optimized supply voltage and frequency; (3) it is easier to achieve load balance among processor cores and to distribute heat across the die; and (4) it can potentially produce lower die temperatures and improve reliability and leakage. However, while ad-hoc methods of selecting few blocks may work based on a designer's experience, this may not work as today's Multi-Processor System-on-Chip (MP-SoC) design which becomes more and more complex. Consequently, System-on-Chip design nowadays needs techniques which can provide an efficient method of enabling a chip to compute complex applications and to fit area-wise on a single chip according to today's technology trends.

1.1.2 Conventional on-Chip Communication Scheme

A communication scheme is composed of an interconnection backbone, physical interfaces, and layered protocols which make the on-chip communication take place among components on an MP-SoC. As the MP-SoC complexity scales up, intra-chip communication requirements are becoming crucial. Data-intensive systems such as multimedia devices, mobile installations, and multi-processor platforms need a flexible and scalable interconnection scheme to handle a huge amount of data transactions on chip. Customarily, dedicated point-to-point wires are adopted as sets of application-specific global on-chip links that connect the top-level modules. However, as wire density and length grow with the system complexity, the communication architecture based on point-to-point wires becomes no more feasible due to its poor scalability and reusability. Specifically, as signals are carried by the global wires across a chip, these metal wires typically do not scale in length with technology. Propagation delay, power dissipation, and reliability will be the serious issues of global wires in deep submicron VLSI technology. According to [2], as silicon technologies advance to 50 nm and beyond, global wires will take 6–10 cycles to propagate, which will then far outweigh gate delays and make cross-chip long wire timing difficult to meet. Keeping track of the status in all elements and managing the global communication among top-level modules by a centralized way are no longer feasible. Therefore, reusable on-chip bus interconnect templates such as ARM's AMBA [3] and IBM's CoreConnect [4] are commonly used in current MP-SoCs design, such that the modules can share the same group of interconnection wires in a bus-based communication architecture.

However, on-chip bus allows only one communication transaction at a time according to the arbitration result, thus the average communication bandwidth of each processing element is in inverse proportion to the total number of IP cores in a system. This character makes a bus-based architecture inherently not scalable for a complex system in today's MP-SoC design. Implementing multiple on-chip buses in a hierarchical architecture or in a separated manner may alleviate this scalability constraint, but it requires application-specific grouping of processing elements and design of different communication protocols to meet the application requirements. Furthermore, whenever a new application needs to be designed for, or a new set of peripherals needs to be added, a chip designed with only simple buses will lack means of efficiently determining feasibility, not to mention optimality [5]. In addition, attempts to guarantee quality-of-service (QoS) for system performance will be a manually intensive task. Therefore, bus-based design needs to be exchanged with a method that is flexible, scalable, and reusable.

1.1.3 Emergence of Network-on-Chip

Since the latest process technology allows for more processors and more cores to be placed on a single chip, the emerging MP-SoC architecture, which demands high throughput, low latency, and reliable global communication services, cannot be met by current dedicated bus-based on-chip communication infrastructure. Trying to achieve such designs with a bus structure could be problematic for a number of reasons including timing closure, performance issues, and scalability. Specifically, as the feature size of modern silicon devices shrinks below 50 nm, global inter-connection delays constrain attainable processing speed. Device parameter variations further complicate the timing and reliability issues. A paradigm shift focusing on communication-centric design, rather than computation-centric design, seems to be the most promising approach to address these communication crises [6–11]. Consequently, in the past few years, a new methodology called Network-on-Chip has been introduced as a means of solving these issues by introducing a structured and scalable communication architecture.

1.2 Concept of Network-on-Chip

Network-on-Chip has been proposed in recent years as a promising solution of on-chip communication network to provide better scalability, performance, and modularity for current MP-SoC architectures [5, 7, 12, 13]. Cross-chip long wires are structured and divided into smaller pieces, thus their electrical properties can be optimized and well controlled. Global signals are transmitted in a pipeline fashion to enhance the operating frequency and to cope with the signal integrity problem in communication. NoC can also promote design productivity by supporting modularity which is convenient to reuse and verify in a higher level of abstraction. Data transmitted across a chip is handled by the intermediate network control units along its route in a distributed manner.

Network-on-chip is a general-purpose on-chip interconnection network that offers great promises to mitigate the ever increasing communication complexity of modern MP-SoC designs. An NoC advocates a communication-centric design style, where a general-purpose communication backbone will first be deployed; then application-specific client logics, such as processors, memory subsystems, peripheral device controllers, etc., will be mapped onto pre-allocated empty slots to form a complete system. This is analogous to the modern land development process where road and communication infrastructures are laid first before specific buildings are designed and built. Along this direction, the city-block style tiled NoC architecture as proposed in [5, 7] has gained high popularity due to its simplicity and flexibility. Mimicking modern city block layout in such architecture, the chip area is divided into rectangular tiles where client logic IPs are placed. The "streets" between tiles are reserved for pre-defined general-purpose on-chip network routing fabrics. Each tile of the 2-D mesh-based on-chip network includes a 5-port router that can transfer data via two unidirectional channels to the local tile, as well as to the north, west, south, and east neighboring routers. In other words, the generic on-chip network is formed by a mesh grid of routers. Each router not only is responsible for the communication needs of its associated tile, but also will route through-traffics originated from and bound toward other tiles.

1.3 Layers in a Network-on-Chip Design

Network-on-Chip is a platform-based interconnection design, which includes vast and complex knowledge from many different scientific fields ranging from the bottom level of solid-state physical interconnection to the uppermost application-software level. Therefore, NoC researches need to be resolved at different levels and can be categorized into physical, network, and application layers based on different abstraction levels. At each partitioning, there exist different problems and solutions.

1.3.1 Physical Layer

The emphasis on physical layer is focused on signal drivers and receivers, as well as design technologies for resorting and pipelining signals on wiring. In addition, as technology advanced to ultra deep submicron (DSM), smaller voltage swings and shrinking feature size translate to decreased noise margin, which cause the on-chip interconnects less immune to noise and increase the chances of non-determinism in the transmission of data over wires (transient fault) [2, 14–17]. Electrical noise due to cross-talk, electromagnetic interference (EMI), and radiation-induced charge injection will likely produce timing error and data errors and make reliable on-chip interconnect hard to achieve.

Error control schemes and utilization of the physical links to achieve reliability are the main concern of this layer. First, a credible fault model must be developed. Then, an error control scheme that is low power, low area, high bandwidth, and low latency must be designed. In NoC design, packet-based data transmission is an efficient way to deal with data errors because the effect of errors is contained by packet boundaries that can be recovered on a packet-by-packet basis.

1.3.2 Network Layer

Network topology or interconnect architecture is an important issue in this layer, which determines how the resources of network are connected, thus refers to the static arrangement of channels and nodes in an interconnection network. Irregular forms of topologies can be derived by mixing different forms of communication architectures in a hierarchical, hybrid, or asymmetric way by clustering partition, which may offer more connectivity and customizability at the cost of complexity and area. In addition, optimization of a topology, which affects the connectivity of the routers and the distance of any one core to the other, is difficult. Furthermore, the tradeoff between generality and customization that respectively facilitate scalability and performance is important. As future designs become more complex, the non-recurring costs of architecting and manufacturing a chip will become more and more expensive. A homogenous NoC is one where the cores and routers are all the same, while a heterogeneous NoC selects individual cores from an IP library and may have its communication architecture customized to suit the needs of an application. Since NoC designs must be flexible enough to cover a certain range of applications, most of the state-of-the-art NoC designs use a mesh or torus topology because of its performance benefits and high degree of scalability for two-dimensional systems, yet it may not achieve the best performance for a single application [13, 18].

In addition, the network layer also needs to deal with the switching and routing data between processing elements. First, packetizing algorithms deal with the decomposition of a message into packets at source nodes and their assembly at destination nodes. Then, the transmission of packets can be executed by the choice of routing algorithms and flow-control methods based on different network topologies [6]. Routing algorithm determines the path strategy of a packet from its source node to the destination node, while flow-control establishes the type of connection among successive nodes on the path. Determining packet routes and resolving conflicts between packets when the same route is requested, with respect to improving on-chip communication performance, are two of the important responsibilities of a router.

Conventional design of a router consists of circuit-switched fabrics and an arbitration controller. In each arbitration decision, more than one path can be constructed by the crossroad switch as long as no contention exists between these paths. For most existing switch designs, virtual-channel flow-control based router design, which provides better flexibility and channel utilization with smaller buffer size, is a well-known technique from the domain of multiprocessor networks [19–26].

1.3.3 Application Layer

At the application layer, target applications will be broken down into a set of computation and communication tasks such that the performance factors like energy and speed can be optimized. Placement of cores on an NoC has to be optimized to reduce the amount of total communication or energy but at the same time recognizing the limitations of any one particular link. The task mapping and communication scheduling problem is an instance of a constrained quadratic assignment problem which was known to be NP-hard [27]. Given a target application described as a set of concurrent tasks with an NoC architecture, the fundamental questions to answer are: (1) how to topologically place the selected set of cores onto the processing elements of the network, and (2) how to take into consideration the complex effects of network condition, which may change dynamically during task execution, such that the metrics of interest are optimized [28]. To get the best tradeoff between power and performance, application mapping and scheduling should be considered with several kinds of architecture parameters.

1.4 Motivation and Contributions

Many researches focused on the improvement of communication efficiency in an NoC have been done. In order to make every application running on an NoC follow the same basic operation principle defined in the backbone interconnection architecture, several issues should be encapsulated into the design. First of all, backbone architecture design should be scalable to the growing amount of traffics in a graceful manner. Also, flexibility is important to deal with the increasing complexity of an application. Various approaches have been explored for improving NoC performance: such as packet routing technique, application mapping and scheduling, topology synthesis, and flow-control [26]. However, considering the physical level of a backbone, the interconnect wire between routers is also an important factor in determining the total performance of a system.

1.4.1 Motivation

In a city-block tiled NoC architecture, neighboring routers are connected via a pair of hard-wired unidirectional communication channels. One link will support out-going traffic and the other link will support in-coming traffic. This kind of conventional NoC architecture has gained high popularity due to its simplicity and flexibility. However, from the simulation results obtained under various traffic conditions, it is often observed that the out-going link may be flooded with out-going traffic while the incoming link remains idle. This leads to performance loss and inefficient resource utilization.

This uneven NoC traffic pattern is very similar to the uneven traffic flow pattern during rush hours on a city highway in a metropolis. A common solution to alleviate such a problem is to implement *reversible lanes* (*counter-flow lanes*) to relieve congestions of the opposing traffic direction. A reversible lane is a highway driving lane with dynamically reversible driving direction assignment. Using electronic signs, the driving direction on a counter-flow lane can be reversed to provide more capacity to the direction with heavier traffic volume.

In this book, similar idea is explored as a mechanism to relieve intermittent traffic congestion in the NoC communication backbone, and hence enhance overall performance. Specifically, a key innovation is proposed to replace the pair of unidirectional links between routers by a pair of bidirectional links that can be dynamically self-configured to facilitate data traffics in either out-going or in-coming direction. This added flexibility promises better bandwidth utilization, lower packet delivery latency, and higher packet consumption rate at each on-chip router.

1.4.2 Contributions

The research topics of this book are generally focused on the network layer which builds the major communication backbone of an NoC. A mechanism similar to *reversible lanes* was proposed to relieve intermittent traffic congestion in an NoC fabric. To facilitate this bidirectional traffic, a novel *inter-router traffic control algorithm* was devised to allow neighboring routers to coordinate the specific directions of the pair of links between them for each data packet. A novel *router architecture* that supports bidirectional links with dynamic self-configuration capability was also proposed.

This novel Bidirectional NoC (BiNoC) router architecture was developed to support dynamic self-reconfiguration of flow direction in each channel, and to enable the most bandwidth utilization between routers. In this BiNoC router, adjacent routers can negotiate the flow directions of connecting channels according to a channel-direction control protocol. This new design concept of a self reconfigurable bidirectional link that enables reconfiguring the transmission direction of a data channel can dynamically adjust the bandwidth based on the real-time traffic requirement. It was shown that the channel-direction control protocol is deadlock-free and starvation-free. Finally, a cycle-accurate behavioral simulator was developed to validate the potential performance gain of our proposed new approach. This simulator is capable of simulating cycle-true traffic behaviors of a moderate size NoC over different traffic patterns. Very encouraging simulation results have been observed as shown in the following chapters. In addition, the implementation detail and experimental results as described in the following chapters proved that this design concept can produce significant performance improvement with reasonable hardware design overhead, making the concept realistic and suitable for NoCs.

In this book, we consider the basic communication backbone design of an NoC and provide a novel BiNoC architecture with a dynamically self-reconfigurable bidirectional channel to break the conventional performance bottleneck caused by bandwidth restriction. Key technical contributions of this work include:

1. A novel BiNoC architecture featuring dynamically self-reconfigured bidirectional channels was proposed. It promises to enhance performance through better resource utilization.
2. The BiNoC router architecture is area-efficient and utilizes smaller buffer size than a conventional unidirectional NoC router while delivering better performance.
3. A new distributed channel-direction control protocol that intelligently and automatically determines the channel transmission direction using local information was devised. It is shown that this protocol is deadlock free and starvation free.
4. The channel-direction control protocol can also intelligently and dynamically reconfigure the channel direction according to the real-time QoS requirement.
5. A new virtual-channel management technique considering QoS requirement was integrated into the BiNoC architecture adequately, where a prioritized routing restriction is applied to limit the best-effort (BE) traffic routing flexibility and leave more communication bandwidth choices for guaranteed-service (GS) traffic flows.
6. A novel NoC fault-tolerant scheme named Bidirectional Fault-Tolerant NoC (BFT-NoC) is proposed to utilize bidirectional channel to provide NoC the fault-tolerance capability instead of detouring packets as in traditional schemes.

7. An efficient power aware task and communication scheduling algorithm was proposed with a unique feature of utilizing the configurability of a bidirectional channel in BiNoC to trade the data transmission time for power expenditure.

1.5 Organization of Book Chapters

The organization of this book is divided into three parts as: "Part I: Introduction to Network-on-Chip," "Part II: Network-on-Chip Design Methodologies Exploration," and "Part III: Case Study: Bi-directional NoC (BiNoC) Architecture." In Part I, the preliminary knowledge of conventional NoC designs and flow-control mechanisms are reviewed in Chap. 2. In Part II, high performance NoC routing techniques are introduced in Chap. 3. Then, the performance-energy trade-off analysis of NoC is carried out in Chap. 4. For energy saving, an energy-aware task scheduling technique for NoC is discussed in Chap. 5. Regarding the major contribution of this book, in Part III, the potential inefficiency of the conventional NoC architecture with a unidirectional channel structure is highlighted in Chap. 6. Thus, a novel BiNoC architecture with flit-buffer flow-control is introduced. A distributed channel-direction control protocol and its detailed operations are described. Afterward, the properties of this proposed bidirectional channel-direction control protocol and BiNoC implementation overhead are analyzed. In Chap. 7, a QoS-aware BiNoC based on the connection-less design concept, which can immediately adjust the inter-router channel direction and assign a higher priority for the critical GS traffic, is depicted. In Chap. 8, a fault tolerant scheme for the BiNoC architecture is provided. Moreover, a novel power-aware task and communication scheduling algorithm for BiNoC is provided to trade the data transmission time for power expenditure in Chap. 9. A brief conclusion is drawn in Chap. 10.

References

1. F. N. Najm, "A Survey of Power Estimation Techniques in VLSI Circuits," *IEEE Transactions on Very Large Scale Integrations Systems*, vol. 2, no. 4, pp. 446–455, December 1994
2. R. Ho, K. W. Mai, and M. A. Horowitz, "The Future of Wires," *Proceedings of the IEEE*, vol. 89, no. 4, pp. 490–504, April 2001
3. ARM, *AMBA Specification Rev* 2.0, ARM Limited, 1999
4. IBM, 32-*bit Processor Local Bus Architecture Specification Version* 2.9, IBM Corporation
5. L. Benini and G. DeMicheli, "Networks on Chips: a New SoC Paradigm," *IEEE Transactions on Computers*, vol. 35, no. 4, pp. 70–78, January 2002
6. W. J. Dally and B. Towles, *Principles and Practices of Interconnection Networks*, Morgan Kaufmann, 2004
7. W. J. Dally and B. Towles, "Route Packets, Not Wires: On-Chip Interconnection Networks," in *Proceedings of the Design Automation Conference*, pp. 684–689, June 2001

8. M. Kistler, M. Perrone, and F. Petrini, "Cell Multiprocessor Communication Network: Built for Speed," *IEEE Micro*, vol. 26, no. 3, pp. 10–23, May 2006
9. L. Seiler, D. Carmean, E. Sprangle, T. Forsyth, P. Dubey, S. Junkins, A. Lake, R. Cavin, R. Espasa, E. Grochowski, T. Juan, M. Abrash, J. Sugerman, and P. Hanrahan, "Larrabee: A Many-Core x86 Architecture for Visual Computing," *IEEE Micro*, vol. 29, no. 1, pp. 10–21, January 2009
10. D. Wentzlaff, P. Griffin, H. Hoffmann, L. Bao, B. Edwards, C. Ramey, M. Mattina, C. C. Miao, J. F. Brown, and A. Agarwal, "On-Chip Interconnection Architecture of the Tile Processor," *IEEE Micro*, vol. 27, no. 5, pp. 15–31, September 2007
11. J. Howard, S. Dighe, Y. Hoskote, S. Vangal, D. Finan, G. Ruhl, D. Jenkins, H. Wilson, N. Borkar, G. Schrom, F. Pailet, S. Jain, T. Jacob, S. Yada, S. Marella, P. Salihundam, V. Erraguntla, M. Konow, M. Riepen, G. Droege, J. Lindemann, M. Gries, T. Apel, K. Henriss, T. L. Larsen, S. Steibl, S. Borkar, V. De, R. Van Der Wijngaart, and T. Mattson, "A 48-Core IA-32 Message-Passing Processor with DVFS in 45nm CMOS," in *Proceedings of the IEEE International Solid-State Circuits Conference*, pp. 108–109, February 2010
12. A. Hemani, A. Jantsch, S. Kumar, A. Postula, J. Oberg, M. Millberg, and D. Lindvist, "Network-on-Chip: an Architecture for Billion Transistor Era," in *Proceedings of the IEEE NorChip Conference*, pp. 1–8, July 2000
13. S. Kumar, A. Jantsch, J. P. Soininen, M. Forsell, M. Millberg, J. Oberg, K. Tiensyrja, and A. Hemani, "A Network-on-Chip Architecture and Design Methodology," in *Proceedings of the International Symposium on Very Large Scale Integration*, pp. 105–112, April 2000
14. R. Hegde and N. R. Shanbhag, "Toward Achieving Energy Efficiency in Presence of Deep Submicron Noise," *IEEE Transactions on Very Large Scale Integration Systems*, vol. 8, no. 4, pp. 379–391, August 2000
15. C. Constantinescu, "Trends and Challenges in VLSI Circuit Reliability," *IEEE Micro*, vol. 23, no. 4, pp. 14–19, July 2003
16. N. Cohen, T. S. Sriram, N. Leland, S. Butler, and R. Flatley, "Soft Error Considerations for Deep-Submicron CMOS Circuit Applications," in *Proceedings of the International Electron Devices Meeting Technical Digest*, pp. 315–318, December 1999
17. P. Shivakumar, M. Kistler, S. Keckler, D. Burger, and L. Alvisi, "Modeling the Effect of Technology Trends on the Soft Error Rate of Combinational Logic," in *Proceeding of the Dependable Systems and Networks*, pp. 389–398, June 2002
18. C. Grecu and M. Jones, "Performance Evaluation and Design Trade-Offs for Network-on-Chip Interconnect Architectures," *IEEE Transactions on Computers*, vol. 54, no. 8, August 2005
19. M. Rahmani, M. Daneshtalab, A. A. Kusha, S. Safari, and M. Pedram, "Forecasting-Based Dynamic Virtual-channels Allocation for Power Optimization of Network-on-Chips," in *Proceedings of the International Conference on VLSI Design*, pp. 151–156, January 2009
20. N Kavaldjiev, G. Smit, and P. Jansen, "A Virtual-channel Router for on-Chip Networks," in *Proceedings of the System-on-Chip Conference*, pp. 289–293, September 2004
21. W. J. Dally, "Virtual Channel Flow Control," *IEEE Transactions on Parallel and Distributed Systems*, vol. 3, no. 2, pp. 194–205, March 1992
22. E. Rijpkema, K. G. W. Goossens, A. Radulescu, J. Dielissen, J. V. Meerbergen, P. Wielage, and E. Waterlander, "Trade-offs in the Design of a Router with Both Guaranteed and Best-Effort Services for Networks-on-Chip," in *Proceedings of the Design Automation and Test in Europe Conference*, pp. 350–355, March 2003
23. H. S. Wang, L. S. Peh, and S. Malik, "A Power Model for Routers: Modeling Alpha 21364 and InfiniBand Routers," *IEEE Micro*, vol. 23, no. 1, 2003
24. R. Mullins, A. West, and S. Moore, "Low-Latency Virtual-Channel Routers for On-Chip Networks," in *Proceedings of the International Symposium on Computer Architecture*, pp. 188–197, June 2004
25. K. Kim, S. J. Lee, K. Lee, and H. J. Yoo, "An Arbitration Look-ahead Scheme for Reducing End-to-End Latency in Networks-on-Chip." in *Proceedings of the International Symposium on Circuits and Systems*, pp. 2357–2360, May 2005

26. P. Guerrier and A. Greiner, "A Generic Architecture for On-Chip Packet-Switched Interconnections," in *Proceedings of the Design Automation and Test in Europe Conference*, pp. 250–256, Mar. 2000
27. M. R. Garey and D. S. Johnson, *Computers and Intractability: a Guide to the Theory of NP-Completeness*, Freeman and Company, 1979
28. R. Marculescu, U. Y. Ogras, L. S. Peh, N. E. Jerger, and Y. Hoskote, "Outstanding Research Problems in NoC Design: System, Microarchitecture, and Circuit Perspectives," *IEEE Transactions on Computer-Aided Design of Integrated Circuits and Systems*, vol. 28, no. 1, pp. 3–21, January 2009

Chapter 2
Preliminaries

Network-on-Chip is the term used to describe an architecture that has maintained readily designable solutions in face of communication-centric trends. In this chapter, we will briefly review some concepts on the design of an NoC router architecture. Various flow-control mechanisms with its corresponding router architecture and design considerations will be included in this chapter.

2.1 Background Knowledge

A typical NoC architecture consists of multiple segments of wires and routers. In a tiled, city-block style of NoC layout, the wires and routers are configured much like street grids of a city, while the client (e.g., logic processor cores) are placed on city blocks separated by wires. A network interface module transforms data packets generated from the client logic (processor cores) into fixed-length flow-control digits (flits). The flits associated with a data packet consist of a header (or head) flit, a tail flit and a number of body flits in between. This array of flits will be routed toward the intended destination in a hop-by-hop manner from one router to its neighboring router.

In a city-block style NoC, each router has five input ports and five output ports corresponding to the north, east, south, and west directions as well as the local processing element. Each port will connect to another port on the neighboring router via a set of physical interconnect wires (channels). The router's function is to route flits entering from each input port to an appropriate output port, and then toward the final destinations. To realize this function, a router is equipped with an input buffer for each input port, a 5×5 cross-bar switch to re-direct traffic to the desired output port, and necessary control logic to ensure correctness of routing results.

Usually, for each data packet, the corresponding head flit specifies its intended destination. After examining the head flit, the router control logic will determine

S.-J. Chen et al., *Reconfigurable Networks-on-Chip*,
DOI: 10.1007/978-1-4419-9341-0_2,

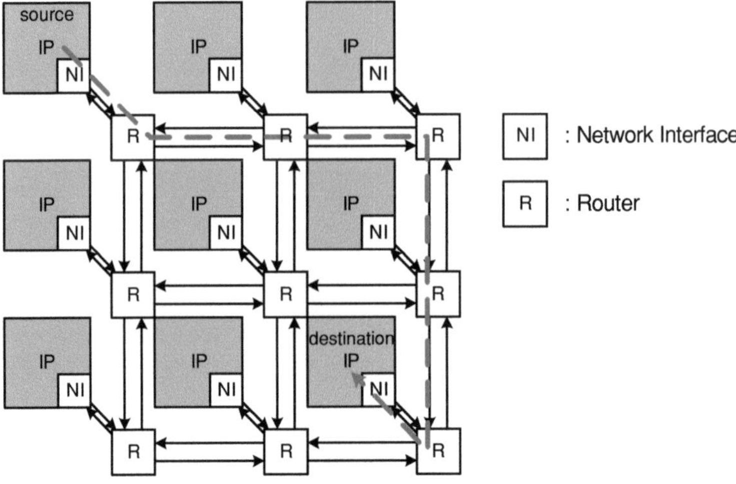

Fig. 2.1 Typical NoC architecture in a mesh topology

which output direction to route all the subsequent (body and tail) flits associated with this data packet according to the routing algorithm applied.

2.2 Conventional Network-on-Chip Architecture

A typical NoC consists of computational processing elements (PEs), network interfaces (NIs), and routers. The latter two comprise the communication architecture.

An NI is used to packetize data before using the router backbone to traverse the NoC. Each PE is attached to an NI that connects the PE to a local router. When a packet was sent from a source PE to a destination PE as shown in Fig. 2.1, the packet is forwarded hop by hop on the network via the decision made by each router. In some NoC architectures that are equipped with error control mechanisms, NIs are also used to encode and decode the data by the error control code applied.

An NoC router is composed of switches, registers, and control logic that collectively perform routing and channel arbitration to guide the flow of packets in the network as illustrated in Fig. 2.2. For each router, the packet is first received and stored in an input buffer. Then the control logic in the router is responsible to make routing decision and channel arbitration. Finally, the granted packet will traverse through a crossbar to the next router, and the process repeats until the packet arrives at its destination.

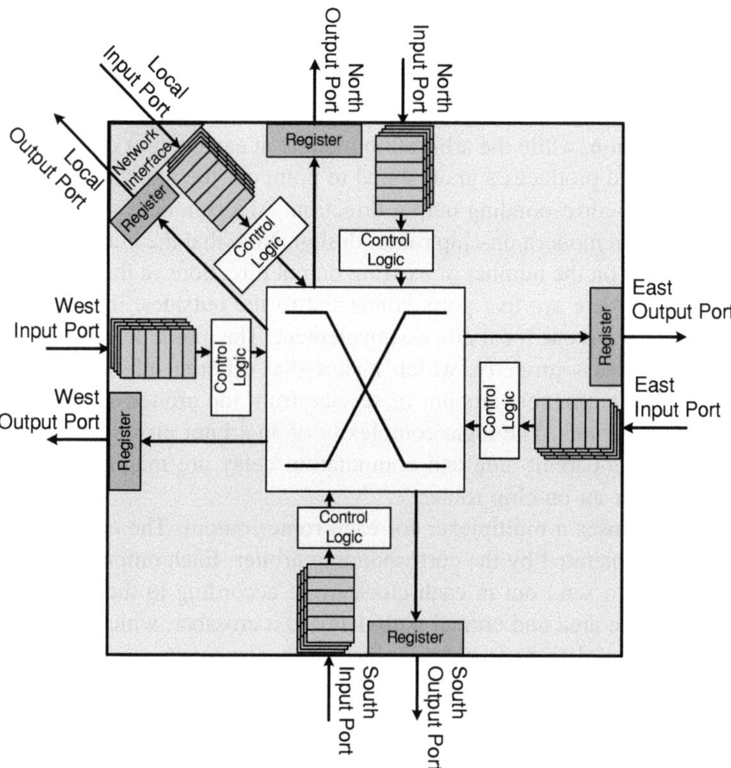

Fig. 2.2 Typical NoC router architecture

2.3 Conventional Router Architecture

Packets delivered by routers are partitioned in a flit-by-flit basis. Each flit of a packet arrives at a router and store in a memory buffer until it can traverse to the next hop of the route. The first flit in the buffer memory will be processed by the control logic to determine whether it is allowed to be forwarded and which output direction it should proceed to. The decision made by the control unit is based on the computation result of routing, arbitration, and the downstream buffer space. After the control setup is done, the flit passes through the crossbar switch to its desired output direction. Since all the decisions are settled by the control logic at the input side, flits will never be stalled at the output ports.

The input buffers in a router design are used to maintain the channel availability until the buffer spaces are exhausted. A larger buffer memory will bring a better throughput and lower latency but however result in large area overhead and power consumption. Usually, a buffer is implemented as a first-in-first-out (FIFO) queue, where data flits can be processed according to the order of their arrival time. Other mechanism such as the dynamic buffer allocation scheme proposed in [1], which trades off complexity with performance, can achieve better buffer space utilization by using a linked-list memory structure.

The two major components in a control logic are the routing and arbitration modules. The routing module processes flits to generate the direction requests at each input direction, while the arbitration module at each output direction receives these requests and produces a grant signal to point out the wining request that can pass through the corresponding output direction. Algorithmic routing with simple logic is popular in modern on-chip router design. Note that the number of direction requests is based on the number of existing output directions in the router. In a 2-D mesh topology, there are five ports connected to the outsides, including the four neighbor routers and one local processing element. The major design concern of an arbiter is its fairness property, which means that different requestors should be provided with a reasonable amount of service from the arbiter according to their individual requirement. The logic complexity of an arbiter grows with the number of input requests. Circuit area and computation delay are major concerns in the arbiter design for an on-chip router.

The crossbar uses a multiplexer for each router output. The control input of a multiplexer is generated by the corresponding arbiter. Each output port can select at most one flit to send out in each clock cycle according to the respective arbitration result. The area and critical path delay of a crossbar, which are affected by the total number of data ports and their bus width, can occupy a significant amount of design concern in a router.

Packet transmissions in wormhole routing are segmented into three flit types: header, body, and tail flits. Head flits are responsible for initiating and reserving the channel bandwidth at each router node for their followers. Body and tail flits are guided by the routes that the header flit has created, and the reserved channel bandwidth will be released by the tail flit to finish the transit of a packet at the current node.

2.4 Flow-Control Mechanism

The performance of NoC communication architecture is dictated by its *flow-control* mechanism. Adding buffers to networks significantly improves the efficiency of a flow-control mechanism since a buffer can decouple the allocation of adjacent channels. Without a buffer, the two channels must be allocated to a packet (or flits) during consecutive cycles, or the packet must be dropped or misrouted [2]. More specifically, with *buffered flow-control*, when a packet arrives at a router, it must first occupy some resources, such as channel bandwidth and buffer capacity, depending on the flow-control methodology. Each router must juggle among multiple input data streams from multiple input ports and route them to appropriate output ports with the highest efficiency.

Buffered flow-control methods can be classified into *packet-buffer flow-control* and *flit-buffer flow-control* based on their granularity of buffer allocation and channel bandwidth allocation [2]. Since allocating resources in unit of flit can achieve more storage utilization efficiency than in unit of packet. Two types of *flit-buffer flow-control* architectures are commonly used in NoC: the *wormhole* flow-control and the *virtual-channel* flow-control.

2.4.1 Packet-Buffer Flow-Control

Packet-buffer flow-control allocates network resources in a packet-by-packet basis. Examples are *store-and-forward flow-control* and *virtual-cut-through flow-control.*

In store-and-forward method, each node must ensure that it has already received and stored an entire packet before forwarding it to the downstream node. While the virtual-cut-through scheme can forward a packet as long as there is enough buffer space to receive a packet at the downstream node. As a result, virtual-cut-through introduces lower communication delay than store-and-forward does. However, packet-buffer flow-control needs larger size of buffer space in one node because of its inefficient use of buffer storage. In addition, allocating channels in units of packets will increase contention latency.

2.4.2 Wormhole Flow-Control Based Router

Wormhole flow-control improves performance through a finer granularity of message allocation at flit-level instead of packet-level. This technique allows more efficient use of buffer than the packet-buffer flow-control mechanism since the buffer size in each router can be reduced significantly [3, 4].

A typical three-stage pipelined NoC router architecture based on wormhole flow-control is shown in Fig. 2.3. Every input port has a FIFO-based input buffer, which can be seen as a single virtual-channel used to hold blocked flits. To facilitate wormhole flow-control based routing [2], the *routing computation* (RC) module will send a channel request signal to the *switch allocator* (SA) for data in each input buffer. If the downstream buffer at a neighboring router has vacant space, SA will allocate the channel and route the data flits through the crossbar switch toward the designated downstream router at the *switch traversal* (ST) stage.

However, wormhole flow-control based switching technique saves buffer size at the expense of throughput since the channel is owned by a packet, but buffers are allocated on a flit-by-flit basis. As such, an idle packet may continue block a channel even when another packet is ready to use the same channel, leading to inefficient resource utilization. This is the well-known *head of line* (HoL) blocking problem. Therefore, virtual-channel flow-control based router architecture was proposed to reduce blocking effect and to improve network latency.

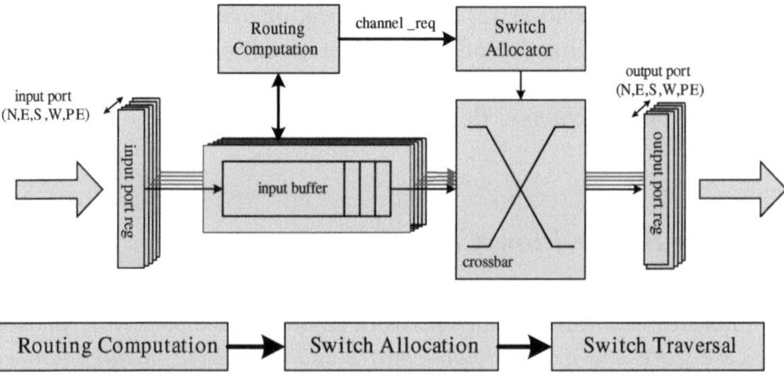

Fig. 2.3 Typical router design based on wormhole flow-control

2.4.3 Virtual-Channel Flow-Control Based Router

Virtual-channel flow-control assigns multiple virtual-paths, each with its own associated buffer queue, to the same physical channel; thus it increases throughput by up to 40% over wormhole flow-control and helps to avoid possible deadlock problems [5–7].

A virtual-channel flow-control router architecture as shown in Fig. 2.4 can be seen as a remedy to the shortcoming of the wormhole flow-control scheme. By multiplexing multiple virtual-channels into the same input buffer, an idle packet will no longer block other packets that are ready to be routed using the shared physical channel. In a typical virtual-channel flow-control based router, the flits are routed via a four-stage pipeline: *routing computation*, *virtual-channel allocation*, *switch allocator*, and *switch traversal.*

One incoming flit that arrives at a router is first written to an appropriate input virtual-channel queue and waits to be processed. When a head flit reaches the top of its virtual-channel buffer queue and enters the RC stage, it is decoded by the RC module and generates an associated direction request. The direction request of this flit is then sent to the VA module to attain virtual-channel at the downstream router. There might be some contentions among packets that request for the same virtual-channel at the downstream router. The loser packets will be stalled at the VA stage and the following flit in the previous stage will also be blocked due to this contention failure. Note that the processes of RC and VA actually take place only on the head flit. The subsequent body flits and tail flit of a packet simply accede to the routing decision acquired by the head flit and require no further processing at the RC and VA stages. Once a decision on the output virtual-channel selection is made at the VA stage, the SA module will assign physical channels to intra-router flits. Flits granted with a physical channel will traverse through the crossbar switch to the input buffer of the down-stream router during the ST stage, and the process repeats until the packet arrives at its destination.

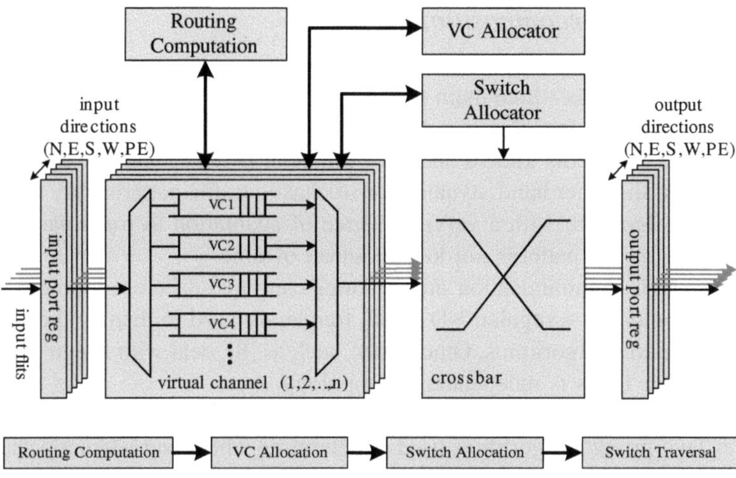

Fig. 2.4 Typical router design based on virtual-channel flow-control

The same as a wormhole flow-control based router, a typical virtual-channel flow-control based router, used by a 2-D mesh-type NoC architecture, contains ten hardwired, physical, and unidirectional data communication channels. Two unidirectional data channels are connected, each in an opposite direction as an input and an output channels, to the neighboring routers. The crossbar, which is a 5 × 5 switch fabric used to connect an input channel to an output channel, in these two typical router architectures can support up to ten unidirectional channels for data transmission. However, the channel bandwidth utilization is not flexible in this hardwired 5-input 5-output NoC router.

For example, while a flit f_A in the west input buffers/virtual-channels is requesting for the north output channel which is being used by another flit f_B from the east input buffer/virtual-channels. f_A has to wait for the north output channel but has no chance to use the other idle channel at the north direction because it is hardwired for input channel, which can only receive data from the neighboring router. However, if all ten unidirectional channels are replaced by bidirectional channels, the channel utilization will be more efficient. Back to the previous example, while the neighboring router is not using the other channel in north direction, the local router can dynamically self-reconfigure this channel as a second output channel and the contention at the north direction can be relieved.

2.5 Routing and Arbitration Techniques

A general problem pertaining to the routing and arbitration algorithms can be stated as: given an application graph which can be represented by a unique traffic pattern, and a communication architecture, find a decision function at each router for selecting an output port that achieves a user-defined objective function.

2.5.1 Problem Decomposition

The above problem has three main parts: a traffic pattern, an NoC communication architecture, and an algorithm which best satisfies a set of user-defined objectives. First, the traffic patterns known ahead of time can be dealt with by a scheduling algorithm. On the other hand, dynamic or stochastic traffic patterns rely on the use of a routing algorithm with a varying degree of adaptation to route packets. Our focus will be on the patterns not known ahead of time.

Second, NoC communication architectures can have different topologies. The most common one is a regular 2-D mesh, frequently used to display the behavior of adaptive routing algorithms. Other work, such as [8], deal with irregular regions in meshes. Our focus is independent of topology.

The third part deals with the algorithms themselves and the objectives to achieve. Two primary algorithms used to determine where and when a packet will move are: routing and arbitration. A routing algorithm decides which direction each input packet should travel. Arbitration is the process of deciding which input packet request should be granted when there are more than one input packet requests for the same output port.

2.5.2 State-of-the-Art

A typical router in an NoC is responsible for moving the received packets from the input buffers, with its routing and arbitration algorithms, to the output ports. The decisions which a router makes are based on the information collected from the network. Centralized decisions refer to making decisions based on the information gathered from the entire network [9]. Distributed decisions refer to making decisions based only on the information generated by the local router or nearby routers. Distributed routing, the focus of this book, allows NoCs to grow in size without worrying about the increasing order of complexity within a centralized routing unit. An example of centralized routing is the AntNet algorithm [10], which depends on global information to make routing decisions, thus needs extra ant buffers, routing tables, and arbitration mechanisms at each node.

There are some distributed routing algorithms which only rely on local information. They have been proposed as being efficient and still maintaining low overhead and high scalability. Routing algorithms in this category include deterministic and adaptive algorithms. Under realistic traffic patterns which pose the problem of hotspot traffic congestion areas, *XY* deterministic routing failed to avoid hotspots and resulted in high average latencies [11]. Adaptive routing guides the router to react to hotspots created by different traffic patterns, by allowing a packet at the input buffer to request more than one output port or direction [12]. While minimal routing algorithms prevent livelock from occurring, adaptive routing introduces the possibility of deadlock, which can be prevented by applying odd–even turn model restrictions to the routing decision [13].

As presented in [11], the DyAD router dynamically switches from deterministic to adaptive routing when congestion is detected, since deterministic routing achieves low packet latency under low packet injection rates. Neighboring nodes send indication to use adaptive routing when their buffers are filled above a preset threshold. Under these conditions, the router dictates that packets are routed in the direction with more available input-buffer slots. This minimal adaptive algorithm, used in the presence of hotspots and increasing congestion rates, pushes back the saturation point of the traffic in the network. Another extension of adaptive routing is the Neighbors-on-Path (NoP) algorithm [14], which allows each router to monitor two hops away the input buffers of the routers in order to detect potential congestion earlier. By earlier detection of the buffer fill level, routes can avoid congestion better. *DyXY* is an algorithm which utilizes a history of buffer fill-levels to make decisions [15]. The algorithms presented in [16, 17] utilize variants of buffer fill-level to make decisions.

In addition to making a routing decision based on the buffer information of downstream packets, the other part of a router's decision making is the arbitration of packets. When multiple input packets are designated to be forwarded to the same next hop destination, arbitration algorithms such as round-robin or first-come first-serve (FCFS) have been proposed to resolve the output port contention. These arbitration algorithms could be designed to relieve upstream buffers with higher congestion. Contention-Aware Input Selection (CAIS) algorithm [18] is an improved arbitration algorithm that contributes to reduce the routing congestion situation by relieving hotspots of upstream traffic, determined by requests from the upstream traffic.

More works have been proposed to deal with some variance of the routing or arbitration algorithms. Sometimes, we categorize the former ones as methods of congestion avoidance; in other words, they evaluate downstream network conditions to avoid sending packets towards the congested areas so as not to aggravate the congestion conditions. We categorize the latter as methods of congestion relief; in other words, they evaluate upstream network conditions to determine which area had the most congestion to send first in order to quickly diffuse the congested situation.

2.6 Quality-of-Service Control

There is a wide range of possibilities for implementing guaranteed services on a network. Referring to the state-of-the-art QoS mechanisms for NoCs, they can be categorized into two types of schemes: connection-oriented (circuit-switching) and connection-less (packet-switching).

2.6.1 Connection-Oriented Scheme

In connection-oriented schemes, guaranteed service packets traverse on some particular channels or buffers that were reserved for them. Specifically, the connection path between the source and destination pair of GS packets is built at the time before they are injected onto the network [19–26]. However, this kind of static pre-allocation may result in high service latency and does not consider hotspots created by temporal shifts in data requirements, thus leads to a rather unscalable NoC.

Connection-oriented QoS mechanism is reliable to achieve QoS requirement, since connections are created guaranteeing tight bounds for specific flows. Two types of the programming models for constructing the setup phase were presented: centralized programming and distributed programming. Centralized programming sets up the reservations by a configuration manager which takes over all the resources in the network. On the contrary, distributed program models let all the resource reservations to be handled by each local router. The centralized method is simpler to achieve while it is only suitable for small-size systems. Despite the hardware overhead in routers, a distributed program model has acquired popularity in a large system because of its better flexibility.

However connection-oriented QoS mechanism comes with greater hardware overhead in control and storage for resource reservations and poor scalability because complexity grows with each node added. Furthermore, bandwidth usage is inefficient and resource allocation has to be considered on a worst case basis. Moreover, the setup phase of guaranteed traffic presents a timing overhead which may result in inefficiency for non-deterministic applications.

2.6.2 Connection-Less Scheme

The connection-less scheme is an alternative way to support different service levels in NoCs where the resource authorities are prioritized according to the QoS requirement of a traffic flow [23]. This is a distributed technique which allows traffic to be classified into different service levels. These service levels can often coincide with different virtual-channels inside the switch. As two traffic flows with different QoS requirements are presented on the same channel simultaneously, the higher prioritized flow can interrupt the lower one and traverse this channel antecedently [23, 27]. It is more adaptive to network traffic and potential hotspots and can better utilize the network.

Different from the connection-oriented schemes, connection-less schemes do not execute any resource reservation. In contrast, multiple traffic flows share the same priority or the same resource thus could cause unpredictable conditions [28]. The traffic with higher service level is guaranteed in a relative fashion in a connection-less scheme by prioritizing each type of traffic flow. However, while the connection-less scheme provides a coarser QoS support as the connection-oriented schemes, they can offer a better adaptation of communication to the varying network traffic. Furthermore, better bandwidth utilization and less hardware cost can be achieved since the traffic is allocated with network resources dynamically. With the consideration of performance requirements for each service level, a network designer can select an appropriate bandwidth implemented in an NoC to both meet the QoS constraints and save the wiring cost [23, 29, 30].

Although connection-oriented communication guarantees tight bounds for several traffic parameters, an erroneous decision of resource reservation might cause an unexpected performance penalty. While in a connection-less network, a non-optimal priority assignment has less degradation of throughput though it provides coarse QoS support. As pointed out in [31], guaranteed services require resource reservation for the worst-case in a connection-oriented, which causes a lot of wasted resource. In addition, some quantitative modeling and comparison of these two schemes, provided in [32], has shown that under a variable-bit-rate application, connection-less technique provides a better performance in terms of the end-to-end packet delay. These comparisons can help to design an application-specific NoC using a suitable QoS scheme.

2.7 Reliability Design

The trend towards constructing large computing systems incorporated with a many-core architecture has resulted in a two-sided relationship involving reliability and fault tolerance consideration. While yield has always been a critical issue in recent high performance circuitry implementation, the document of International Technology Roadmap for Semiconductor (ITRS) [33] states that "Relaxing the requirement of 100% correctness for devices and interconnects may dramatically reduce costs of manufacturing, verification and test." The general principle of fault tolerance for any system can be divided in two categories:

1. Employment of hardware redundancy to hide the effect of faults.
2. Self-identification of source of failure and compensate the effect by appropriate mechanism.

If we can make such a strategy work, a system will be capable of testing and reconfiguring itself, allowing it to work reliably throughout its lifetime.

2.7.1 Failure Types in NoC

Scaling chips however increase the probability of faults. Faults to be considered in an NoC architecture can be categorized into permanent (hard-fault) and transient fault (soft-fault) [34, 35]. The former one reflects irreversible physical changes, such as electro-migration of conductor, broken wires, dielectric breakdowns, etc. In this case, permanent damages in a circuit cannot be repaired after manufacture. Therefore, the module which is suffering a permanent fault should turn off its function and inform neighboring modules of this information. Then, re-routing packets with an alternative path will be re-calculated deterministically or dynamically according to the need. However, this may induce non-minimal path routing and increase the complexity of routing decision. Hardware redundancy such as spare wire or reconfigurable circuitry can also be used to avoid using of faulty modules [36–39]. In the latter case, several phenomena, such as neutron and alpha particles, supply voltage swing, and interconnect noise, induce the packet invalid or misrouted. Usually, a transient fault is modeled with a probability of bit error rate under an adequate fault model. In an NoC system, intra-router or inter-router functionality errors may happen, to understand how to deal with the most common sources of failures in an NoC, Park et al. provided comprehensive fault tolerant solutions relevant to all stages of decision making in an NoC router [40].

2.7.2 Reliability Design in NoC

A number of fault tolerant methods were proposed in [41, 42] for large-scale communication systems. Unfortunately, these algorithms are not suitable for an NoC, because they will induce significant area and resource overhead. Dumitras et al. proposed a flood-based routing algorithm for NoC, named *stochastic communication*, which is derived from the fault tolerance mechanism used in the computer network and distributed database fields. Such stochastic-communication algorithm separates computation from communication and provides fault tolerance to on-chip failures [33, 43]. However, to eliminate the high communication overhead of flood-based fault tolerance algorithm, Pirretti et al. promoted a redundant random-walk algorithm which can significantly reduce the overhead while maintaining a useful level of fault tolerance [44]. However, the basic idea of sending redundant information via multipath to achieve fault tolerance may cause much higher traffic load in the network, and the probabilistic broadcast characteristic may also result in additional unpredictable behavior on network loading.

Therefore, in a distributed NoC router considering practical hardware implementation, the error control scheme used to detect/correct inter-router transient fault in an NoC is required to have smaller area and shorter timing delay. An error control code that adapts to different degrees of detection and correction and has a low timing overhead will ease its integration into a router. The fault-tolerant method utilizing error detection requires an additional retransmission-buffer specially designed for NoCs when the errors are detected. Error control schemes, such as the Reed-Solomon code proposed by Hoffman et al. have been used on NoCs [45]. But as their results show, the long delay would degrade the overall timing and performance of an NoC Router.

2.8 Energy-Aware Task Scheduling

The availability of many cores on the same chip promises a high level of parallelism to expedite the execution of computation-intensive applications. To do so, a program must first be represented by a *task graph* where each node is a coarse-grained task (e.g., a procedure or a sub-routine). Often, a task needs to forward its intermediate results to another task for further processing. This inter-task data-dependency is represented by a directed arc from the origin task to the destination task in the task graph. Tasks that have no inter-task data dependency among themselves can be assigned to multiple processor cores to execute concurrently. As such, the total execution time can be significantly shortened.

A *real-time application* is an application which execution time must be smaller than a *deadline*. Otherwise, the computation will be deemed a failure. To implement an application on an MC-NoC platform for parallel execution, each task in the task graph will be *assigned* to a processor core. Depending on the city-block distance between two tiles, inter-task communication will take different amount of communication delay. For a particular application, proper task assignment will reduce communication delay while maximizing parallelism such that the total execution time can be minimized. For a real-time application, if the total execution time is less than the pre-defined deadline of the application, the *slacks* between them could be exploited to reduce energy consumption.

The execution time of a task may vary depending on the clock frequency the processor core is running. One technique to adjust the clock frequency of individual time on an MC-NoC is *Dynamic Voltage Scaling* (DVS). When the clock frequency slows down, often the associated energy consumed by a running task is also reduced. Hence, in addition to assigning tasks to the processor cores located at appropriate tiles, another design objective would be to use DVS to save some energy while conforming to the deadline constraint, with perhaps smaller slacks.

Previously, it has been shown that the minimum energy multi-processor task scheduling problem is NP-hard [46–48]. For real-time applications, it was proposed that execution of some tasks can be slowed down using DVS on corresponding tiles without violating the deadline timing constraint [49]. Several DVS-enabled uni-processors have been implemented. Test results running real-world applications showed significant power saving up to 10 times [50]. For multi-processor core systems implemented to execute a set of real-time dependent tasks, Schmitz et al. [51–53] presented an iterative synthesis approach for DVS-enabled processing element based on genetic algorithms (GA). They proposed a heuristic PV-DVS algorithm specifically for solving the voltage scaling. Kianzad et al. improved the previous work by combining assignment, scheduling, and power management in a single GA algorithm [54]. However, GA-based design optimization suffers slow convergence and lower desired quality. Chang et al. [55] proposed using Ant Colony Optimization (ACO) algorithm. Common to these approaches is that when PV-DVS is applied for power reduction, it is applied to one task (tile) at a time and is done after assignment and scheduling. Zhang et al.

[56] and Varatkar et al. [57] proposed using a list scheduling algorithm to find an initial task schedule, and the DVS problem was solved by integer linear programming. The idea behind these methods is to maximize the available slack in a schedule so as to enlarge the solution space of using DVS. However, the communication infrastructures used in these works are either a point-to-point interconnect, or a bus architecture. Hu et al. [58] proposed an energy-aware scheduling (EAS) algorithm that considers the communication delay on an NoC architecture. However, DVS frequency adjustment was not considered.

References

1. Y. Tamir and G. L. Frazier, "Dynamically-Allocated Multi-Queue Buffers for VLSI Communication Switches," *IEEE Transactions on Computers*, vol. 41, no. 6, pp. 725–737, June 1992
2. W. J. Dally and B. Towles, *Principles and Practices of Interconnection Networks*, Morgan Kaufmann, 2004
3. W. J. Dally and C. L. Seitz, "The Torus Routing Chip," *Journal of Distributed Computing*, vol. 1, no. 4, pp. 187–196, January 1986
4. P. Kermani and L. Kleinrock, "Virtual Cut-Through: A New Computer Communication Switching Technique," *Computer Networks*, vol. 3, no. 4, pp. 267–286, September 1979
5. W. J. Dally, "Virtual Channel Flow Control," *IEEE Transactions on Parallel and Distributed Systems*, vol. 3, no. 2, pp. 194–205, March 1992
6. L. S. Peh and W. J. Dally, "A Delay Model for Router Microarchitectures," *IEEE Micro*, vol. 21, no. 1, pp.26–34, January 2001
7. W. J. Dally and C. L. Seitz, "Deadlock-Free Message Routing in Multiprocessor Interconnection Networks," *IEEE Transactions on Computers*, vol. C-36, no. 5, pp. 547–553, May 1987
8. E. Bolotin, I. Cidon, R. Ginosar, and A. Kolodny, "Routing Table Minimization for Irregular Mesh NoCs," *in Proceedings of the Design Automation and Test in Europe Conference*, pp. 1–6, April 2007
9. M. A. Yazdi, M. Modarressi, and H. S. Azad, "A Load-Balanced Routing Scheme for NoC-Based System-on-Chip," *in Proceedings of the Workshop on Hardware and Software Implementation and Control of Distributed MEMS*, pp. 72–77, June 2010
10. M. Daneshtalab, A. A. Kusha, A. Sobhani, Z. Navabi, M. D. Mottaghi, and O. Fatemi, "Ant Colony Based Routing Architecture for Minimizing Hot Spots in NOCs," *in Proceedings of the Annual Symposium on Integrated Circuits and System Design*, pp. 56–61, September 2006
11. J. Hu and R. Marculescu, "DyAD–Smart Routing for Networks-on-Chip," *in Proceedings of the Design Automation Conference*, pp. 260–263, June 2004
12. C. J. Glass and L.M. Ni, "The Turn Model for Adaptive Routing," *Journal of ACM*, vol. 41, no. 5, pp. 874-902, September 1994
13. G. M. Chiu, "The Odd-Even Turn Model for Adaptive Routing," *IEEE Transactions on Parallel and Distributed Systems*, vol. 11, no. 7, pp. 729–738, July 2000
14. G. Ascia, V. Catania, M. Palesi, and D. Patti, "Neighbors On-Path: A New Selection Strategy for On-Chip Networks," *in Proceedings of the IEEE Workshop on Embedded Systems for Real Time Multimedia*, pp. 79–84, October 2006
15. M. Li, Q.A. Zeng, and W. B. Jone, "DyXY - a Proximity Congestion-Aware Deadlock-Free Dynamic Routing Method for Network on Chip," *in Proceedings of the Design Automation Conference*, pp.849–852, July 2006

16. E. Nilsson, M. Millberg, J. Oberg, and A. Jantsch, "Load Distribution with the Proximity Congestion Awareness in a Network-on-Chip," *in Proceedings of the Design Automation and Test in Europe Conference*, pp.1126–1127, December 2003
17. J. Kim, D. Park, T. Theocharides, N. Vijaykrishnan, and C. R. Das, "A Low Latency Router Supporting Adaptivity for on-Chip Interconnects," *in Proceedings of the Design Automation Conference*, pp. 559–564, June 2005
18. D. Wu, B. M. Al-Hashimi, and M. T. Schmitz, "Improving Routing Efficiency for Network-on-Chip through Contention-Aware Input Selection," *in Proceedings of the Asia and South Pacific Design Automation Conference*, pp. 36–41, January 2006
19. M. Millberg, E. Nilsson, R. Thid and A. Jantsch, "Guaranteed Bandwidth using Looped Containers in Temporally Disjoint Networks within the Nostrum Network on Chip," *in Proceedings of the Design Automation and Test in Europe Conference*, pp. 890–895, February 2004
20. K. Goossens, J. Dielissen, and A. Radulescu, "The Æthereal Network on Chip: Concepts, Architectures, and Implementations," *IEEE Design & Test of Computers*, vol. 22, no. 5, pp. 414–421, October 2005
21. P. Vellanki, N. Banerjee, and K. S. Chatha, "Quality-of-Service and Error Control Techniques for Mesh-Based Network-on-Chip Architectures," *ACM Very Large Scale Integration Journal*, vol. 38, no. 3, pp. 353–382, January 2005
22. N. Kavaldjiev, G. J. M. Smit, P. G. Jansen, and P. T. Wolkotte, "A Virtual-channel Network-on-Chip for GT and BE Traffic," *in Proceedings of the Annual Symposium on Emerging VLSI Technologies and Architectures*, pp. 211–216, March 2006
23. E. Bolotin, I, Cidon, R. Ginosar, and A. Kolodny, "QNoC: QoS Architecture and Design Process for Network-on-Chip," *Elsevier Journal of System Architecture*, vol. 50, no.2–3, pp. 105–128, February 2004
24. M. Dall'osso, G. Biccari, L. Giovannini, D. Bertozzi, and L. Benini, "Xpipes: a Latency Insensitive Parameterized Network-on-Chip Architecture for Multiprocessor SoCs," *in Proceedings of the International Conference on Computer Design*, pp. 536–539, October 2003
25. D. Bertozzi and L. Benini, "Xpipes: a Network-on-Chip Architecture for Gigascale System-on-Chip," *IEEE Circuits and Systems Magazine*, vol. 4, no. 2, pp. 18–31, April 2004
26. T. Bjerregaard and J. Sparso, "A Router Architecture for Connection-Oriented Service Guarantees in the MANGO Clockless Network-on-Chip," *in Proceedings of the Design Automation and Test in Europe Conference*, pp. 1226–1231, March 2005
27. M. D. Harmanci, N. P. Escudero, Y. Leblebici, and P. Ienne, "Providing QoS to Connection-Less Packet-Switched NoC by Implementing DiffServ Functionalities," *in Proceedings of the International Symposium on System-on-Chip*, pp. 37–40, November 2004
28. A. Mello, L. Tedesco, N. Calazans, and F. Moraes, "Evaluation of Current QoS Mechanisms in Networks-on-Chip," *in Proceedings of the International Symposium on System-on-Chip*, pp. 1–4, November 2006
29. Z. Guz, E. Bolotin, I. Cidon, R. Ginosar, and A. Kolodny, "Efficient Link Capacity and QoS Design for Network-on-Chip," *in Proceedings of the Design Automation and Test in Europe Conference*, pp. 1–6, March 2006
30. P. Vellanki, N. Banerjee, and K. S. Chatha, "Quality-of-Service and Error Control Techniques for Network-on-Chip Architecture," *in Proceedings of the Great Lakes Symposium on VLSI*, pp. 45–50, April 2004
31. E. Rijpkema, K. G. W. Goossens, A. Radulescu, J. Dielissen, J. V. Meerbergen, P. Wielage, and E. Waterlander, "Trade-offs in the Design of a Router with Both Guaranteed and Best-Effort Services for Networks-on-Chip," *in Proceedings of the Design Automation and Test in Europe Conference*, pp. 350–355, March 2003
32. M. D. Harmanci, N. P. Escudero, Y. Leblebici, and P. Ienne, "Quantitative Modeling and Comparison of Communication Schemes to Guarantee Quality-of-Service in Networks-on-Chip," *in Proceedings of the International Symposium on Circuits and Systems*, pp. 1782–1785, May 2005

33. P. Bogdan, T. Dumitras, and R. Marculescu, "Stochastic Communication: A New Paradigm for Fault Tolerant Networks on Chip," *VLSI Design*, vol. 2007, Article ID 95348, pp. 1–17, 2007
34. R. Marculescu, U. Y. Ogras, L. S. Peh, N. E. Jerger, and Y. Hoskote, "Outstanding Research Problems in NoC Design: System, Microarchitecture, and Circuit Perspectives," *IEEE Transactions on Computer-Aided Design of Integrated Circuits and Systems*, vol. 28, no. 1, pp. 3–21, January 2009
35. M. Ali, M. Welzl, and S. Hessler, "A Fault Tolerant Mechanism for Handling Permanent and Transient Failures in Network on Chip," *in Proceeding of the International Conference on Information Technology*, pp.1027–1032, April 2007
36. M. Yang, T. Li, Y. Jiang, and Y. Yang, "Fault-Tolerant Routing Schemes in RDT(2,2,1)/α-Based Interconnection Network for Network-on-Chip Designs," *in Proceedings of the International Symposium on Pervasive Systems, Algorithms and Networks*, pp. 1–6, December 2005
37. T. Lehtonen, P. Liljeberg, and J. Plosila, "Online Reconfigurable Self-Timed Links for Fault Tolerant NoC," *VLSI Design*, vol. 2007, Article ID 94676, pp. 1–13, 2007
38. K. Kariniemi and J. Nurmi, "Fault Tolerant XGFT Network on Chip for Multi-Processor System on Chip Circuit," *in Proceedings of the International Conference on Field Programmable Logic and Applications*, pp. 203–210, August 2005
39. T. Schonwald, J. Zimmermann, O. Bringmann, and W. Rosentiel, "Fully Adaptive Fault-Tolerant Routing Algorithm for Network-on-Chip Architectures," *in Proceedings of the Euromicro Conference on Digital System Design*, pp. 527–534, August 2007
40. D. Park, C. Nicopoulos, J. Kim, N. Vijaykrishnan, and C. R. Das, "Exploring Fault-Tolerant Network-on-Chip Architectures," *in Proceedings of the Annual IEEE/IFIP International Conference on Dependable Systems and Networks*, pp. 93–104, June 2006
41. Y. Hatanaka, M. Nakamura, Y. Kakuda, and T. Kikuno, "A Synthesis Method for Fault-Tolerant and Flexible Multipath Routing Protocols," *in Proceedings of the International Conference on Engineering of Complex Computer Systems*, pp. 96–105, September 1997
42. W. Stallings, *Data and Computer Communications*, Prentice Hall, 2007
43. T. Dumitras, S. Kerner, and R. Marculescu, "Towards On-Chip Fault-Tolerant Communication," *in Proceedings of the Asia and South Pacific Design Automation Conference*, pp.225–232, January 2003
44. M. Pirretti, G. M. Link, R. R. Brooks, N. Vijaykrishnan, M. Kandemir, and M. J. Irwin, "Fault Tolerant Algorithms for Network-on-Chip Interconnect," *in Proceedings of the IEEE Computer Society Annual Symposium on VLSI*, pp. 46–51, February 2004
45. J. Hoffman, D.A. Ilitzky, A. Chun, and A. Chapyzhenka, "Architecture of the Scalable Communications Core," *in Proceedings of the International Symposium on Networks-on-Chip*, pp. 40–52, May 2007
46. E. S. H. Hou, N. Ansari, and H. Ren, "A Genetic Algorithm for Multiprocessor Scheduling," *IEEE Transactions on Parallel and Distributed Systems*, vol. 5, no. 2, pp. 113–120, February 1994
47. C. M. Krishna and K. G. Shin, *Real-Time Systems*, WCB/McGraw Hill, 1997
48. H. El-Rewini, H. H. Ali, and T. Lewis, "Task Scheduling in Multiprocessor Systems," *IEEE Computer*, vol. 28, no. 12, pp. 27–37, December 1995
49. T. Burd and R. W. Brodersen, "Energy Efficient CMOS Microprocessor Design," *in Proceeding of the Hawaii International Conference on System Sciences*, pp. 288–297, January 1995
50. G. Quan and X. Hu, "Energy Efficient Fixed-Priority Scheduling for Real-Time Systems on Voltage Variable Processors," *in Proceedings of the Design Automation Conference*, pp. 828–833, June 2001
51. M. T. Schmitz and B. M. Al-Hashimi, "Considering Power Variations of DVS Processing Elements for Energy Minimization in Distributed Systems," *in Proceedings of the International Symposium on Systems Synthesis*, pp. 250–255, October 2001

52. M. T. Schmitz, B. M. Al-Hashimi, and P. Eles, "Energy-Efficient Mapping and Scheduling for DVS Enabled Distributed Embedded Systems," *in Proceedings of the Conference on Design, Automation and Test in Europe*, pp 514–521, March 2002
53. M. T. Schmitz, B. M. Al-Hashimi, and P. Eles, "Iterative Schedule Optimization for Voltage Scalable Distributed Embedded Systems," ACM TECS, vol. 3, no. 1, pp. 182-217, February 2004
54. V. Kianzad, S. S. Bhattacharyya, and G. Qu, "CASPER: An Integrated Energy-Driven Approach for Task Graph Scheduling on Distributed Embedded Systems," *in Proceedings of the International Conference on Application-Specific Systems, Architectures and Processors*, pp. 191–197, July 2005
55. P. C. Chang, I. W. Wu, J. J. Shann, and C. P. Chung, "ETAHM: An Energy-Aware Task Allocation Algorithm for Heterogeneous Multiprocessor," *in Proceedings of the Design Automation Conference*, pp. 776–779 June 2008
56. Y. Zhang, X. Hu, and D. Z. Chen, "Task Scheduling and Voltage Selection for Energy Minimization" *in Proceedings of the Design Automation Conference*, pp. 183–188, June 2002
57. G. Varatkar and R. Marculescu, "Communication-Aware Task Scheduling and Voltage Selection for Total Systems Energy Minimization," *in Proceedings of the International Conference on Computer-Aided Design*, pp. 510–517, November 2003
58. J. Hu and R. Marculescu, "Energy-Aware Communication and Task Scheduling for Network-on-Chip Architectures under Real-Time Constraints," *in Proceedings of the Design, Automation and Test in Europe Conference and Exhibition*, pp. 234–239, February 2004

Part II
Network-on-Chips Design Methodologies Exploration

Chapter 3
Techniques for High Performance Noc Routing

In an NoC, designing an efficient routing mechanism is critical to the performance. One crucial issue in the routing strategies is, under the premise of deadlock and livelock freedoms, how to enhance routing adaptivity in order to come up with a flexible and efficient use of the available routing resources. Key NoC performance metrics include low packet delivery latency and high throughput rate. These requirements are critically impacted by the underlying routing algorithm [1].

3.1 NoC Routing Basics

In NoC-based communication systems, the data packets are generally broken into a contiguous sequence of flow units known as flits. Transmitting a packet from a source to its destination requires the sequential transmission of multiple flits over the same path. The path is chosen distributedly by applying the same routing algorithm at each router encountered by the packets en route to their destinations. In practice, the design of an efficient routing mechanism is essential in optimizing the performance of NoC-based communication systems [1, 2].

3.1.1 Characterization of NoC Routing

NoC routing algorithms utilize various strategies in selecting suitable paths. For example, they may always choose a pre-determined path between each particular pair of source and destination nodes (deterministic), or may dynamically choose different paths depending on the traffic conditions (adaptive), or may choose among a set of alternative paths which obey certain routing rules (partially adaptive). Furthermore, NoC routing algorithms may be constrained to the use of only profitable routes (minimal), or may be permitted to select detour routes (non-minimal) for congestion avoidance or fault tolerance purposes.

S.-J. Chen et al., *Reconfigurable Networks-on-Chip*,
DOI: 10.1007/978-1-4419-9341-0_3,

Any routing scheme can have its pros and cons. For example, in minimal path routing, the path length is equal to the two-dimensional city block (mesh) distance between the source and the destination. The constraint of using only minimal paths has the advantages in guaranteeing livelock free and minimal hops for a packet traversal, which simplifies the design of a deadlock-free routing algorithm. However, the minimal routing limitation may lead to performance degradation and function loss in certain conditions. As shown in Fig. 3.1a, the minimal routing path may carry a heavy traffic load, causing an excessive delay (latency) for any flits which must traverse its links. By contrast, non-minimal routing paths offer alternative light-traffic routes which provide additional link bandwidth and therefore reduce the overall latency by avoiding contention as illustrated in Fig. 3.1b. Furthermore, as shown in Figs. 3.1c, d, non-minimal routing provides a significantly improved fault-tolerance performance. However, a non-minimal routing scheme probably requires more implementation overheads (e.g., cost and performance) compared with the minimal one.

3.1.2 Deadlock and Livelock Issues

In any routing algorithm, it is essential to avoid both deadlock and livelock. Deadlock is an anomalous network state in which a circular hold-and-wait dependency relation is formed among the network resources; causing the routing of the packets to be indefinitely postponed as shown in Fig. 3.2. Meanwhile, in livelock, a packet travels continuously around the network without ever reaching its destination since the requested channels are constantly occupied by other packets. Livelock occurs only in adaptive and non-minimal routing. Thus, in contrast to deadlock, livelock is relatively easily avoided by using minimal paths or by imposing certain rules on the non-minimal path selection process [3], such as a probabilistic way [4], and a priority mechanism [5].

3.1.3 Deadlock-Free Routing Schemes in NoCs

Deadlock is by far one of the most difficult problems to solve in NoC designs [3]. Existing NoC routing algorithms utilize either one of the following strategies to handle the deadlock condition, namely *deadlock avoidance* or *deadlock recovery*. Deadlock avoidance schemes impose additional constraints on the routing algorithm such that deadlock can never occur. Meanwhile, the deadlock recovery approach detects and remedies the damage caused by deadlock, then continues normal routing operations.

As shown in Fig. 3.3, deadlock freedom is commonly accomplished using some forms of turn-model based routing algorithms. Of the various turn models which have been proposed, the Odd–Even model [6] is one of the most elaborate, and its

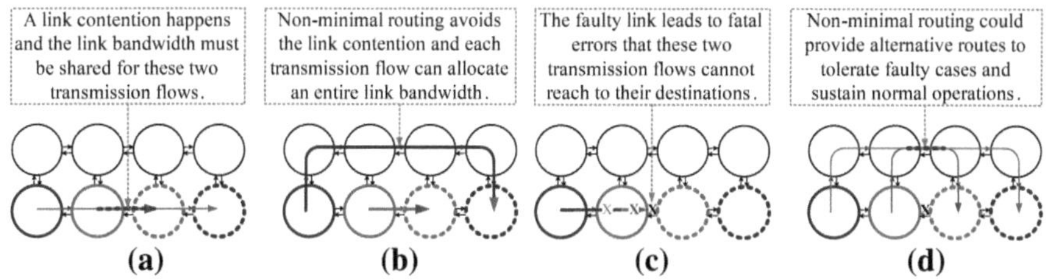

Fig. 3.1 Local congestion scenarios in **a** minimal routing and **b** non-minimal routing; and faulty link cases in **c** minimal routing and **d** non-minimal routing

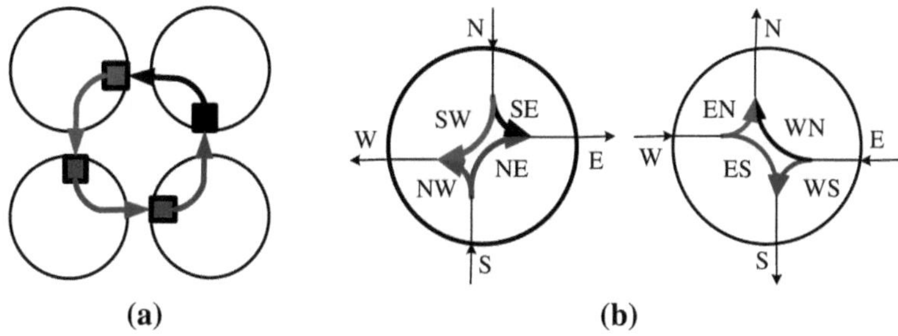

Fig. 3.2 **a** Deadlock condition and **b** eight turn types in a two-dimensional mesh

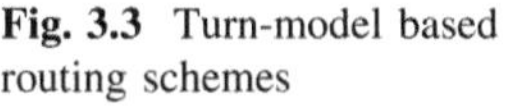
Fig. 3.3 Turn-model based routing schemes

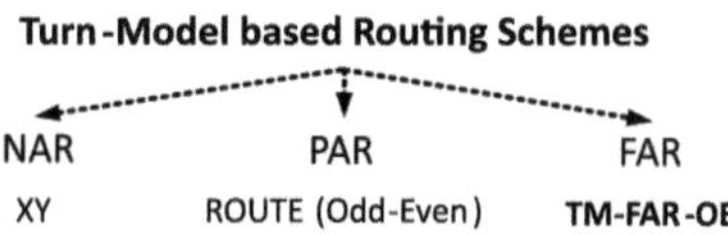

associated minimal routing algorithm, ROUTE, has been extensively applied in NoCs (e.g., DyAD [7], NoP [8], Schafer et al. [9], Lin et al. [10], and Wu [11]). In general, the turn-model based routing algorithms have a lower implementation complexity and a more flexible routing performance than those deadlock-free approaches, such as Virtual Channel (VC) based methods [12, 13], deflection routing algorithms [14, 15], and deadlock recovery approaches [16, 17].

However, traditional turn-model based routing schemes are either Non-Adaptive Routing (NAR) such as XY or Partially Adaptive Routing (PAR) such as Odd–Even [6], which can only use a subset of minimal paths in their classes. In Fully Adaptive Routing (FAR), packets can be routed using all the shortest paths. Routing algorithms usually use VCs to achieve the goal of FAR. However, using extra VCs leads to power and performance penalties in switching packets [6, 18]. For high performance on-chip communications, it is desirable to have a routing algorithm that is fully adaptive, performance-efficient, and deadlock-free. To achieve this goal, a Turn-Model based Fully-Adaptive-Routing (TM-FAR) scheme is proposed in this chapter.

Table 3.1 Routing algorithms, turn models, and prohibited turns

Routing algorithm	Turn model name	Prohibited turn
XY	N/A[a]	NW, SE, NE, SW
West-First	West-First	NW, SW
North-Last	North-Last	NW, NE
Negative-First	Negative-First	NW, ES
ROUTE	Odd–Even	NW, SW in odd column EN, ES in even column

Note[a] : No particular turn model name. N, E, S, W represent north, east, south, west, respectively

3.2 Turn Model Based Routing Basics

NoC routing algorithms utilize the rules specified within the adopted turn model to route the packets toward their destination in such a way that prohibited turns are avoided and the packets do not become stalled. Table 3.1 summarizes the major turn-model based minimal length routing algorithms currently available for deadlock and livelock avoidance. As shown in each algorithm, the packets are routed to their destinations without using certain turns.

3.2.1 Odd–Even Turn Model

This section reviews the turn rules in the Odd–Even turn model [6] and describes the routing criteria applied in ROUTE, the corresponding minimal routing algorithm. The Odd–Even turn model is governed by the following turn rules:

1. *Turn Rule* 1: No packet is allowed to make an EN turn at any router located in an even column, or an NW turn at any router located in an odd column.
2. *Turn Rule* 2: No packet is allowed to make an ES turn at any router located in an even column, or an SW turn at any router located in an odd column.
3. *Turn Rule* 3: (Derived from Theorem 1 of [6]). No packet is allowed to make a 180-degree turn at any router.

Referring to the above rules, any packet is not allowed to take an EN or ES turn at any node located in an even column, and it is not allowed to take an NW or SW turn at any node located in an odd column. Deadlock freedom can be proved, because the rightmost column segment always is short of an essential turn to form a circular waiting path. We demonstrate the principle in Fig. 3.4.

3.2.2 Odd–Even Turn-Model Based Routing Algorithm, ROUTE

In designing any Odd–Even turn-model based routing algorithm, the routing criteria must be consistent with the rules specified by the Odd–Even model. For example, the Minimal Routing (MinR) Criteria in ROUTE [6] are as follows:

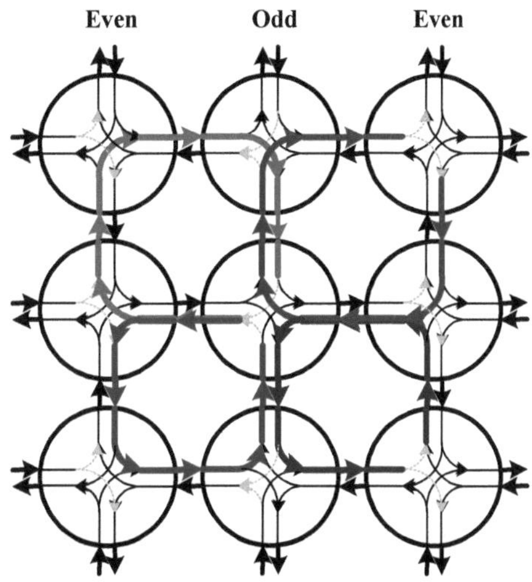

Fig. 3.4 Four transmission examples following the Odd–Even turn model

1. *MinR Criterion* 1: No packet may move in a direction away from its destination (i.e., path selection is constrained to minimal routes).
2. *MinR Criterion* 2: If the destination of a packet is to the west of its source, the packet may not move north or south at any intermediate routers residing in an odd column unless the destination is located in the same column (see Fig. 3.5a).
3. *MinR Criterion* 3: If the destination of a packet is to the east of its source and is located in an even column, the packet must finish routing in the north or south direction before it reaches the column in which the destination is located (see Fig. 3.5b).

The Odd–Even model is regarded as the current state-of-the-art turn model since it does not prohibit any certain turn at all positions, and therefore has a higher degree of routing adaptivity than other turn models [6]. As a result, the ROUTE algorithm (Fig. 4 of [6]) is more elaborate than other routing algorithms such as West-First, North-Last, and Negative-First [19].

3.2.3 Motivations of our Proposed Turn Model Based Routing Schemes

Turn models (i.e., XY and Odd–Even) are the most prevalent methodologies adopted for deadlock-free packet routing in NoCs. However, the developed researches such as the path-selection strategy for adaptive routing [7] and the fault-tolerant routing [10] are highly constrained in the inherent turn prohibitions. Nevertheless, these prohibitions of turn models can be relieved as presented in the next two sections.

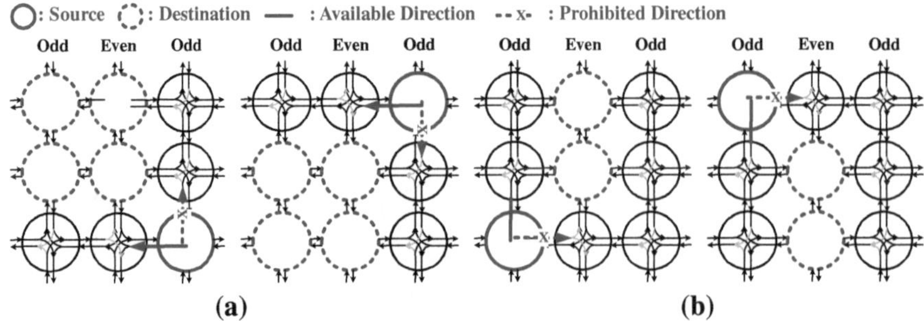

Fig. 3.5 **a** Minimal routing criterion 2 and **b** minimal routing criterion 3

3.3 Proposed Turn-Model Based Fully Adaptive Routing

To relieve the Partially Adaptive Routing (PAR) limitation of the traditional turn-model based routing, a novel Turn-Model based Fully-Adaptive-Routing (TM-FAR) algorithm is proposed. TM-FAR retains the deadlock-free property of traditional turn-model based routing algorithms (e.g., XY, Odd–Even), while alleviating restrictions on turn and path selections. Just like the current Virtual-Channel based Fully-Adaptive-Routing (VC-FAR) algorithm, TM-FAR allows full exploitation of all available minimal paths, yet TM-FAR does not use virtual channels. This fully adaptive routing capability of TM-FAR promises an improved routing adaptivity and an enhanced level of fault-tolerance.

3.3.1 Turn Prohibitions Release

Virtual-Cut-Through (VCT) [20] is a basic packet switching scheme. When the space for the entire packet is available in the target node, VCT can start to forward flits of the received packet. VCT is performance inefficient compared with wormhole and the buffer size needs to be at least equal to the maximal packet.

Essentially, our TM-FAR-OE algorithm uses wormhole switching, follows all the Odd–Even turn rules [6], and does not strictly restrict the size of buffers. Particularly, borrowing concepts from VCT, TM-FAR-OE can release turn prohibitions imposed by Odd–Even in the case where the empty space of an attached input buffer in neighbor nodes is equal to or greater than the size of the packet going to be relayed. Practically, the buffer status of a neighbor node can be checked through the counter for credit-based flow control, and the variable packet length can be recorded in the header flit. We named this technique *Enhanced Virtual-Cut-Through* (EVCT) and depicted the principle in Fig. 3.6.

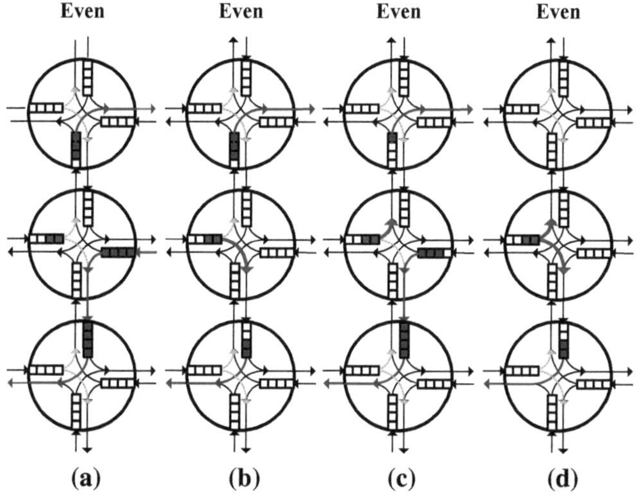

Fig. 3.6 Examples of EN and ES turn prohibitions imposed by Odd–Even are active in **a**; but released in **b**, **c**, and **d** by using EVCT

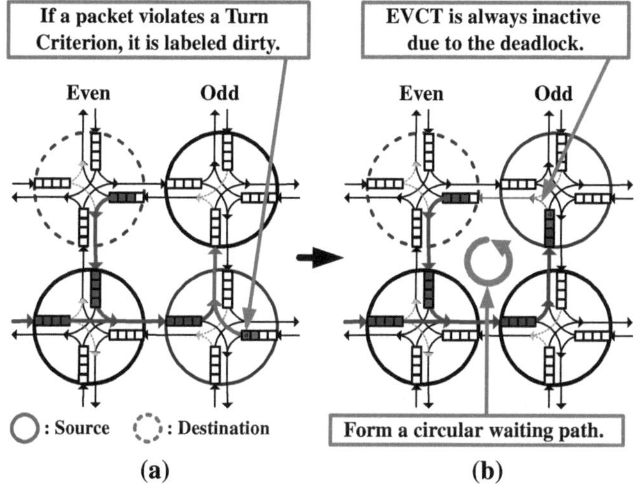

Fig. 3.7 **a** Violated turn criterion case and **b** circular waiting path

3.3.2 *Path Prohibitions Release*

Since the turn prohibitions can be removed by EVCT, we start to consider the possibility of removing the constraints in turn criteria as described above. Unfortunately, deadlocks could exist if we apply EVCT without regarding the turn criteria. A scenario is shown in Fig. 3.7. Following all the turn rules and criteria in Odd–Even,

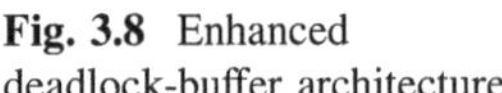

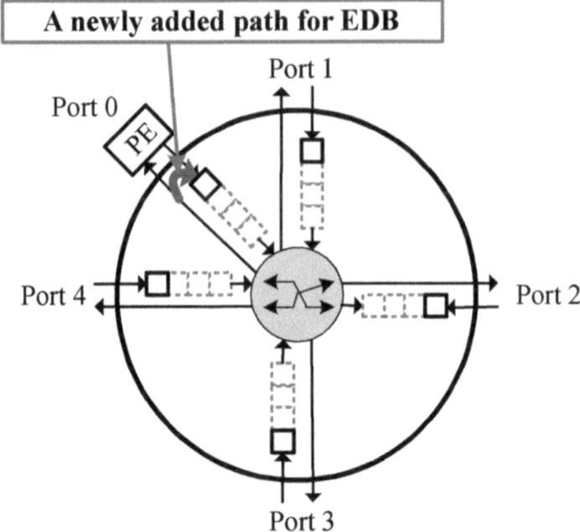

Fig. 3.8 Enhanced deadlock-buffer architecture

EVCT could validate additional turns from the inherent illegal turns of Odd–Even. That is, these additional turns are under certain circumstances, not deadlock-free guaranteed. Figure 3.7b shows that when an Odd–Even prohibited turn becomes the only available path for a packet to reach its destination, a deadlock could be incurred.

To achieve the FAR goal in our algorithm, we designed another technique named *Enhanced Deadlock-Buffer* (EDB). Deadlock-Buffer (DB) is a deadlock recovery scheme proposed in DISHA [16]. DB is a buffer dedicated for when a deadlock is presumed, and DB requires connecting with all input buffers and output ports in a router. Thus, directly applying DB in a chip increases the connection complexity of crossbars and degrades the layout utilization of routers.

To utilize the functionality of DB in our NoC router, in contrast with DISHA's allocating additional buffers as DBs [16], we assign the input buffer for Processing Elements (PEs) another role of DB. Besides, DB is exercised as a deadlock-avoidance mechanism, not deadlock-recovery as in DISHA [16]. We illustrate the updated router in Fig. 3.8. Here, we list five operation rules for EDB:

1. *Buffer Size Requirement*: The size of EDB (i.e., the input buffer of Port 0) is required to be not less than the size of the maximal packet that permits violating a turn criterion.
2. *Dirty Packet Label*: When a packet violates a turn criterion, the router labels it dirty using one bit in the packet header. This bit notices the router that the receiving packet needs to use a prohibited turn to reach its destination. The dirty label can be removed after the packet passes a prohibited turn.
3. *Mutual Exclusion*: The EDB accepts a dirty packet when its empty space is equal to or greater than the dirty packet. Input ports of the router need to obtain a mutual exclusive access grant to EDB before they allow a dirty packet to be entered into its input buffer. In other words, at any one time, there is only one dirty packet existent in the router.
4. *Parking Regulation*: When a router receives a dirty packet, and it cannot forward the packet to the next node because there is no available valid path;

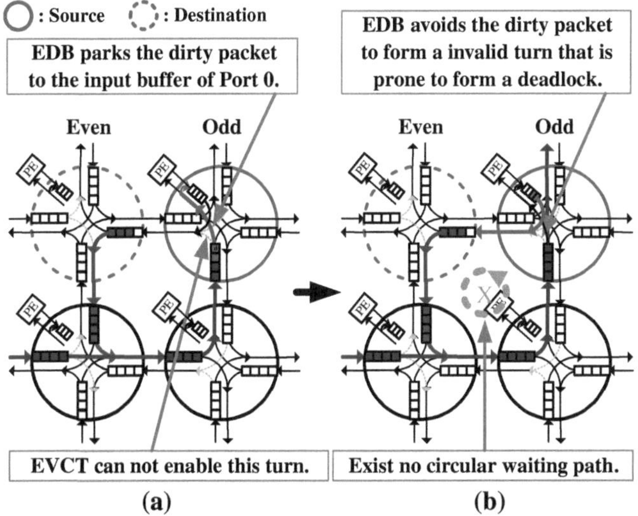

Fig. 3.9 **a** EDB parks the dirty packet in the input buffer of port 0, and **b** the packet can be relayed to the destination without creating an NW turn

immediately, the router shall park this dirty packet to EDB to prevent blocking other normal packets from entering the input buffer. The parked dirty packet leaves EDB to a next node whenever a path is available.

5. *Turn Criteria Violation Principle* (optional): A router is suggested to violate the turn criteria by forwarding packets to an illegal path of Odd–Even only when the input buffer connecting to the valid path is full or crowded.

In Fig. 3.9, the deadlock condition in Fig. 3.7b can be prevented when the routers incorporate the technique of EDB.

3.3.3 Deadlock Freedom and Livelock Freedom

On one hand, EVCT can increase the routing adaptivity, but slightly enhance the routing performance because the path selections are still limited by the turn criteria that may reduce the possibility of activating EVCT. On the other hand, EDB removes the turn and path prohibitions imposed by Odd–Even, the goal of fully adaptive routing can be achieved; however, the performance is decreased, because it takes times to allow dirty packets go through the prohibited turns. But, EVCT and EDB could be a good couple. Since EDB removes turn criteria of Odd–Even, EVCT can be active for each packet in all routers. Through the help of EVCT, dirty packets can speedily pass a prohibited turn validated by EVCT, thus EDB can just be used to buffer the dirty packet that cannot be instantly relayed to a neighbor node and reduce the probability of activating EDB.

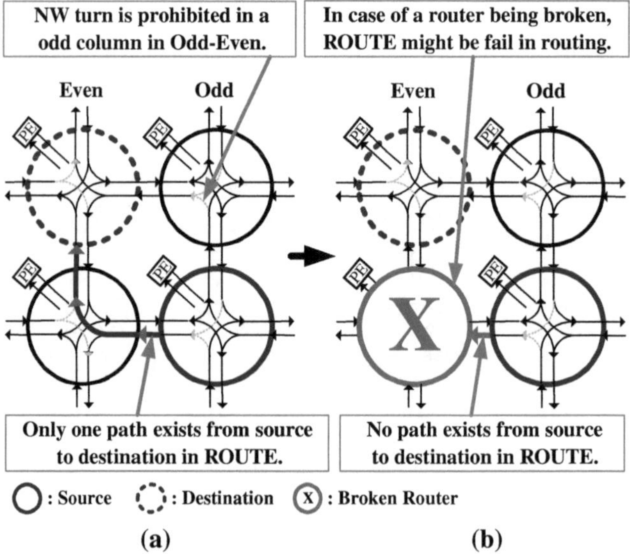

Fig. 3.10 Routing examples of **a** a normal case and **b** a faulty case of ROUTE

TM-FAR-OE is a routing algorithm integrating both EVCT and EDB with Odd–Even. First, TM-FAR-OE is livelock free due to its minimal routing property. Second, TM-FAR-OE guarantees deadlock freedom because the five operation rules of EDB guarantee that a motionless dirty packet can always be buffered in a deadlock buffer.

Theorem *The TM-FAR based routing algorithm that follows the rules adopted in turn model is deadlock-free as long as the180-degree turns are prohibited in a mesh network and all dirty packets in input buffers are in a moving state.*

Proof We prove the theorem by contradiction. Assume that there exists a set of packets $p1, p2, \ldots, pn$, that are deadlocked. Thus, the associated waiting path forms a circular path. Since the 180-degree turns are prohibited, the circular path must include four different, either clockwise or counterclockwise 90-degree turns. In accordance with the adopted turn model, one of these four turns is prohibited with normal packets; therefore, the prohibited turn must be formed by a dirty packet being blocked in an input buffer. Thus, contradiction arises, and we prove the theorem. □

3.3.4 Fault Tolerance Advantage

As a *Partially Adaptive-Routing* (PAR) algorithm, ROUTE [6] probably provides only one of the available minimal paths between a source and destination pair due

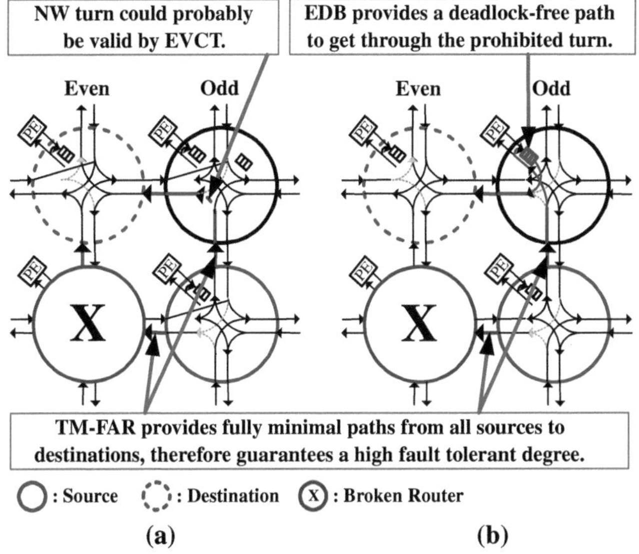

Fig. 3.11 Routing examples of packets getting through the prohibited turn by **a** EVCT and **b** EDB under a faulty case of TM-FAR

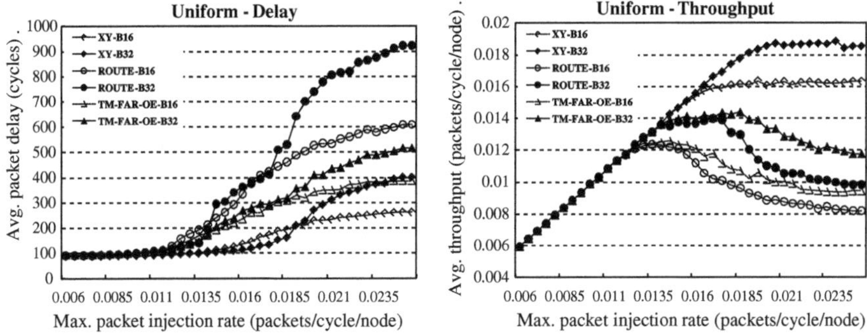

Fig. 3.12 Performance variations in delay and throughput under uniform traffic

to the turn and path prohibitions of Odd–Even. This limitation could cause problem under a possible faulty case as shown in Fig. 3.10.

Considering fault-tolerance will be highly desirable in future on-chip micro-network designs [21]. Some researches [11, 22] extended turn models to achieve a certain degree of fault-tolerance; nevertheless, the proposed methods are highly limited in the inherent turn and path restrictions of a turn model. But now, TM-FAR gets rid of the turn and path prohibitions of a turn model as illustrated in Fig. 3.11. As such, TM-FAR provides additional feasibility to design an algorithm with a higher degree of fault-tolerance and better routing efficiency.

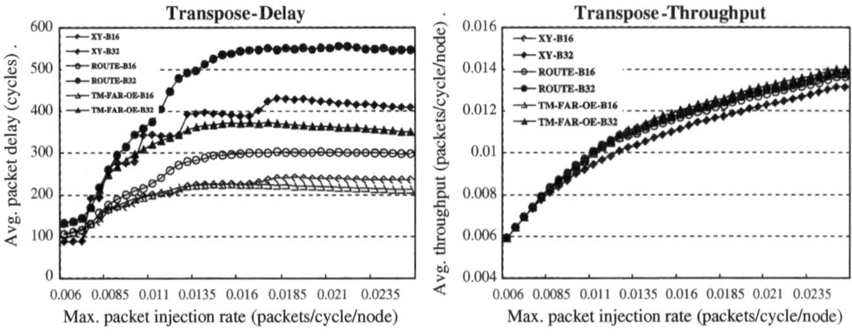

Fig. 3.13 Performance variations in delay and throughput under transpose traffic

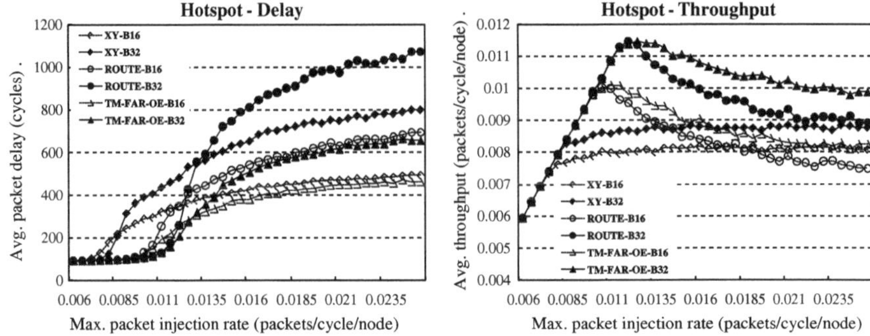

Fig. 3.14 Performance variations in delay and throughput under hotspot traffic

3.3.5 Performance Evaluation

Comprehensive simulations were run in Register Transfer Level using Cadence NC-Verilog. Performance metrics were carried out on an (8×8) mesh network. Each link bandwidth was set to one flit per cycle. Packets were generated and received by a host model attached to Port 0 of each router. In different evaluations, the sizes of buffers were configured in 16 and 32 flits, respectively; the size of packets was randomly distributed between 4 and 16 flits in all simulations. In uniform traffic, a node transmits a packet to any other node with equal probability. In transpose traffic, a node at (i, j) always sends packets to a node at (j, i). In hotspot traffic, uniform traffic is applied, but 20% of packets change their destination to one of the following four selected nodes [(7, 2), (7, 3), (7, 4), (7, 5)] with equal probability. For each traffic load value, the results of packet latency and throughput are averaged over 60,000 packets after the unstable warm-up session of 30,000 arrival packets.

Table 3.2 Performance comparisons among different traffics with **a** buffer sizes of 16 flits and **b** 32 flits

(a)

Buffer Size = 16 flits (16 × 32bits)

Algorithm\Traffic	Average delay			Average throughput		
	Uniform	Transpose	Hotspot	Uniform	Transpose	Hotspot
XY	158.08	208.38	376.00	0.01336	0.01063	0.00792
ROUTE	328.45	259.03	447.28	0.00956	0.01107	0.00824
TM-FAR-OE	232.65	202.55	321.03	0.01018	0.01118	0.00863

Algorithm\Traffic	Normalized by ROUTE			Normalized by ROUTE		
	Uniform (%)	Transpose (%)	Hotspot (%)	Uniform (%)	Transpose (%)	Hotspot (%)
XY	48.13	80.45	84.06	139.65	96.07	96.12
ROUTE	100.00	100.00	100.00	100.00	100.00	100.00
TM-FAR-OE	70.83	78.20	71.77	106.41	100.97	104.72

(b)

Buffer Size = 32 flits (32 × 32bits)

Algorithm\Traffic	Average delay			Average throughput		
	Uniform	Transpose	Hotspot	Uniform	Transpose	Hotspot
XY	182.70	353.65	564.70	0.01432	0.01063	0.00848
ROUTE	423.85	454.33	636.73	0.01092	0.01121	0.00938
TM-FAR-OE	266.80	324.68	409.55	0.01179	0.01132	0.00995

Algorithm\Traffic	Normalized by ROUTE			Normalized by ROUTE		
	Uniform (%)	Transpose (%)	Hotspot (%)	Uniform (%)	Transpose (%)	Hotspot (%)
XY	43.10	77.48	88.69	131.15	94.88	90.39
ROUTE	100.00	100.00	100.00	100.00	100.00	100.00
TM-FAR-OE	62.95	71.46	64.32	108.02	101.05	106.06

3.3.5.1 Effects of Fully Adaptive Routing

In Figs. 3.12, 3.13, and 3.14, when the simulation results of XY routing algorithm were generated under a size of buffers configured in 16 flits, we named it XY-B16. Likewise, XY-B32 reflects the XY routing performed upon a buffer size of 32 flits.

Uniform Traffic: Figure 3.12 shows the results obtained under uniform traffic. We observe that the XY algorithm performed the best. Identical results were shown in [6]. Since the non-adaptive XY embodies global and long-term information for the uniform traffic pattern, it happens to spread traffic much more evenly across paths of a mesh. However, except for XY, TM-FAR-OE achieved

Table 3.3 Performance comparisons among different traffics and buffer sizes

Maximal throughput						
Traffic\Algorithm	16 flits (16 × 32bits)			32 flits (32 × 32bits)		
	XY	ROUTE	TM-FAR-OE	XY	ROUTE	TM-FAR-OE
Uniform	0.01636	0.01234	0.01259	0.01888	0.01399	0.01440
Transpose	0.01317	0.01364	0.01378	0.01316	0.01383	0.01400
Hotspot	0.00820	0.01004	0.01010	0.00883	0.01148	0.01148
Algorithm\Traffic	Maximal throughput normalized by buffer size = 16 flits (16 × 32 bits)					
	XY (%)	ROUTE (%)	TM-FAR-OE (%)	XY (%)	ROUTE (%)	TM-FAR-OE (%)
Uniform	100.00	100.00	100.00	115.36	113.38	114.37
Transpose	100.00	100.00	100.00	99.94	101.44	101.63
Hotspot	100.00	100.00	100.00	107.71	114.38	113.72

the highest saturation point in throughput and performed better than ROUTE in both delay and throughput. Compared with ROUTE by averaging the data of Uniform in Tables 3.2a, b), TM-FAR-OE improved 33.11% in delay and 7.21% in throughput.

Transpose Traffic: The transpose traffic pattern is a kind of specific operations identical to the Matrix-Transpose used in [19]. Figure 3.13 shows that, before the saturation point in throughput and under an identical packet injection rate, TM-FAR-OE and ROUTE provide similar performance in throughput; however, TM-FAR-OE greatly overcomes ROUTE in the packet delay. Compared with ROUTE by averaging the data of Transpose in Tables 3.2a, b, TM-FAR-OE improved 25.17% in delay and 1.01% in throughput.

Hotspot Traffic: Hotspot is a more realistic traffic scenario [6]. Hotspot traffic causes early saturation for all routing schemes due to uneven traffic loads in the network. In contrast to ROUTE that provides partial adaptivity due to the turn and path prohibitions, TM-FAR-OE supports full adaptivity for packets to route around local traffic jams. Figure 3.14 shows that ROUTE and TM-FAR-OE had similar peak throughput value. Besides that, TM-FAR-OE outperformed ROUTE in both performance metrics. Compared with ROUTE by averaging the data of Hotspot in Tables 3.2a, b, TM-FAR-OE improved 31.95% in delay and 5.39% in throughput.

3.3.5.2 Effects of Buffer Size

We also compared the buffer size effects on network performance between one (16 flits) and double (32 flits) sizes of a maximal packet as shown in Fig. 3.12. Referring to Table 3.3, the maximal average throughput in the transpose traffic was improved about 1.01% on average by enlarging the size of the buffers. Except for transpose, the maximal average throughput was enhanced about 14.37% in uniform and 11.93% in hotspot traffic patterns. According to the experimental results as shown in [23], our network performance can also be effectively enhanced by allocating each buffer in multiples of a packet size.

3.4 Remarks

We presented a Turn-Model based Fully-Adaptive-Routing scheme, TM-FAR, that guaranteed deadlock and livelock freedom. In our proposed routing algorithm, TM-FAR-OE, the turn rules and criteria of Odd–Even were no longer being used to prohibit available turns and limit path selections in an NoC router. Instead, we transformed these intrinsic restrictions of turn model into the operation conditions which guided the router to operate in a Virtual-Cut-Through mode or to use Deadlock-Buffers. Compared with state-of-the-art routing algorithm, TM-FAR-OE achieved an averaged delay reduction of 30.08% and throughput rate increase of 4.54%.

References

1. T. Bjerregaard and S. Mahadevan, "A Survey of Research and Practices of Network-on-Chip," ACM Computing Surveys, vol. 38, no. 1, pp. 1-51, March 2006.
2. C. Grecu and M. Jones, "Performance Evaluation and Design Trade-Offs for Network-on-Chip Interconnect Architectures," IEEE Transactions on Computers, vol. 54, no. 8, August 2005.
3. J. Duato, S. Yalamanchili, and L. Ni, Interconnection Networks: An Engineering Approach, Morgan Kaufmann, 2002.
4. S. Konstantinidou and L. Snyder, "The Chaos Router," IEEE Transactions on Computers, vol. 43, no. 12, pp. 1386-1397, December 1994.
5. J. T. Brassil and R. L. Cruz, "Bounds on Maximum Delay in Networks with Deflection Routing," IEEE Transactions on Parallel and Distributed Systems, vol. 6, no. 7, pp. 724-732, July 1995.
6. G. M. Chiu, "The Odd-Even Turn Model for Adaptive Routing," IEEE Transactions on Parallel and Distributed Systems, vol. 11, no. 7, pp. 729-738, July 2000.
7. J. Hu and R. Marculescu, "DyAD–Smart Routing for Networks-on-Chip," in Proceedings of the Design Automation Conference, pp. 260-263, June 2004.
8. G. Ascia, V. Catania, M. Palesi, and D. Patti, "Implementation and Analysis of a New Selection Strategy for Adaptive Routing in Networks-on-Chip," IEEE Transactions on Computers, vol. 57, no. 6, pp. 809-820, June 2008.
9. M. K. F. Schafer, T. Hollstein, H. Zimmer, and M. Glesner, "Deadlock-Free Routing and Component Placement for Irregular Mesh-based Networks-on-Chip," in Proceedings of the International Conference on Computer-Aided Design, pp. 238-245, November 2005.
10. S. Y. Lin, C. H. Huang, C. H. Chao, K. H. Huang, and A. Y. Wu, "Traffic-Balanced Routing Algorithm for Irregular Mesh-Based On-Chip Networks," IEEE Transactions on Computers, vol. 57, no. 9, pp. 1156-1168, September 2008.
11. J. Wu, "A Fault-Tolerant and Deadlock-Free Routing Protocol in 2D Meshes Based on Odd-Even Turn Model," IEEE Transactions on Computers, vol. 52, no. 9, pp. 1154-1169, September 2003.

12. W. J. Dally and C. L. Seitz, "Deadlock-Free Message Routing in Multiprocessor Interconnection Networks," IEEE Transactions on Computers, vol. C-36, no. 5, pp. 547-553, May 1987.
13. L. Schwiebert and D. N. Jayasimha, "Optimal Fully Adaptive Wormhole Routing for Meshes," in Proceedings of the Conference on Supercomputing, pp. 782-791, November 1993.
14. P. Baran, "On Distributed Communications Networks," IEEE Transactions on Communications Systems, vol. 12, no. 1, pp. 1-9, March 1964.
15. T. Moscibroda and O. Mutlu, "A Case for Bufferless Routing in On-Chip Networks," in Proceedings of the International Conference on Computer Architecture, pp. 196-207, January 2009.
16. K. V. Anjan and T. M. Pinkston, "An Efficient, Fully Adaptive Deadlock Recovery Scheme: DISHA," in Proceedings of the Annual International Symposium on Computer Architecture, pp. 201-210, June 1995.
17. Y. H. Song and T. M. Pinkston, "Distributed Resolution of Network Congestion and Potential Deadlock Using Reservation-Based Scheduling," IEEE Transactions on Parallel and Distributed Systems, vol. 16, no. 8, pp. 686-701, August 2005.
18. K. Aoyama and A. A. Chien, "The Cost of Adaptivity and Virtual Lanes in a Wormhole Router," VLSI Design, vol. 2, no. 4, pp. 315-333, January 1995.
19. C. J. Glass and L.M. Ni, "The Turn Model for Adaptive Routing," Journal of ACM, vol. 41, no. 5, pp. 874-902, September 1994.
20. P. Kermani and L. Kleinrock, "Virtual Cut-Through: A New Computer Communication Switching Technique," Computer Networks, vol. 3, no. 4, pp. 267-286, September 1979.
21. L. Benini and G. DeMicheli, "Networks on Chips: a New SoC Paradigm," IEEE Transactions on Computers, vol. 35, no. 4, pp. 70-78, January 2002.
22. Z. Zhen, A. Greiner, and S. Taktak, "A Reconfigurable Routing Algorithm for a Fault-tolerant 2D-Mesh Network-on-Chip," in Proceedings of the Design Automation Conference, pp. 441-446, June 2008.
23. J. Hu, U. Y. Ogras, and R. Marculescu, "System-Level Buffer Allocation for Application-Specific Networks-on-Chip Router Design," IEEE Transactions on Computer-Aided Design of Integrated Circuits and Systems, vol. 25, no.12, pp. 2919-2933, December 2006.

Chapter 4
Performance-Energy Tradeoffs for Noc Reliability

The NoC architecture promises reliable high performance low power on-chip communication. To realize such promises, performance-energy trade-off analysis is carried out in this chapter to compare two competing error control strategies: forward error correction (FEC) versus automatic re-transmission request (ARQ). Contrary to previously reported results, we show that the ARQ scheme would consume more power than the FEC scheme to offer the same level of reliability when the power consumption of the re-transmission buffers is factored into the equation. This new finding leads to the conclusion that FEC error control strategy is more suitable for NoC implementation compared to ARQ.

4.1 Reliability in NoC

The latest process technology allows more processors and more cores to be placed on a single chip. Extensive researches of communication centric design on how to integrate the reusable IPs into a single chip become an important issue [1, 2]. In other words, the design of communication architecture plays a major part in the whole system's performance, throughput, and reliability in such SoC system. However, as technology advanced to ultra deep submicron (DSM), small voltage swings and shrinking feature size translate to decreased noise margin, which caused the on-chip interconnects to be less immune to noise and increased the chances of non-determinism in the transmission of data over wires (transient fault) [3–7]. Electrical noise due to cross-talk, electromagnetic interference (EMI), and radiation-induced charge injection will be likely to produce timing error and data errors and make reliable on-chip interconnect hard to achieve.

In addition to higher susceptibility to transient fault, trends in CMOS technology also result in higher power consumption on global on chip wire. Global wire delay model in DSM shows that the on chip global communication requires increasingly higher energy consumption [8]. Hence, designing low-power

S.-J. Chen et al., *Reconfigurable Networks-on-Chip*,
DOI: 10.1007/978-1-4419-9341-0_4, © Springer Science+Business Media, LLC 2012

interconnect circuit is a challenging problem because it compels us to address the issues of energy reduction and reliable operation in a unified manner. More specifically, since voltage swing is reduced in lower power design, the relative effect of noise becomes larger and renders a chip more sensitive to environment noise.

To mitigate the reliability problem, the bus based system usually uses an error detection scheme cooperated with a retransmission protocol. However, bus structure that cannot be scaled up with the increasing system complexity becomes a bottleneck for a number of reasons, including timing closures, performance issues, and design reuse. Therefore, NoC has been introduced as a means of solving these issues by introducing structured and scalable communication architecture in the past few years. The packetized communication on NoC makes it easier to transmit data incorporated with error control information. Also, breaking a transfer wire into modular segments in NoC design also allows error control to be implemented on a per segment basis. Modular design for reliability is in line with scalability concerns when designing communications for complex chips. But on chip communication differs from wide-area network, the fault tolerant method required evaluation and optimization in terms of area, delay, and power consumption trade-off to improve system reliability [9–17].

To design a reliable NoC architecture, the issues of energy consumption and impact of traffic load need to be considered, but have not been fully considered in existing works. Different reliability enhancement schemes would consume different amount of hardware resources with different level of performance impacts such as energy, traffic delay and throughput. Therefore, it is necessary to devise an in-depth analysis and comprehensive experiment to explore the design space of different NoC reliability enhancement strategies. The outcome of this chapter would (1) offer a cost-benefit comparison framework of different NoC reliability schemes so that future methods can be compared with existing methods under the same set of objective criteria; (2) provide a guideline on how Error Correction Code (*ECC*) measures should be chosen under different design constraints.

The solution for keeping error rate under acceptable margins may use error correcting code and/or parity-based detection with retransmission scheme built into the router or operation at higher voltage to increase the signal to noise ratio. As such in this chapter, we present a comprehensive analysis of the energy performance trade-offs among several different reliability enhancement schemes. We define a new power consumptions measurement metric in terms of energy per useful bit (E_{pub}), and propose ways to estimate data throughput and impact of traffic latency of different reliability enhancement schemes under commonly used single event upset (SEU) fault model. To facilitate fair comparison, we set common reliability constraints such as mean time to failure (*MTTF*) of system and residual error probability (*REP*) when analyzing different error control schemes.

Bertezzi et al. provided an energy efficiency tradeoff of the FEC and ARQ for on chip communication links and reported that under the same system reliability, ARQ consumes less energy than FEC [18]. Ejlali et al. provided a simultaneous consideration of fault-tolerance, energy-efficiency, and performance; they reported that ARQ is the most preferable choice at low noise power and proved that the hybrid ARQ/FEC is more advantageous as the noise power increases [15]. However, these researchers have not considered the overhead of the retransmission buffer. This is non-neglectable since the retransmission buffer costs even more energy than the encoder or decoder of an error control scheme. Even though the error probability is low, the retransmission still needs to be active to hold the data in case of any errors. Comparing to existing works, this chapter has made several tangible contributions:

1. We consider more aspects of performance and energy metrics for better encapsulating the tradeoffs made in different low-overhead NoC error control schemes include FEC, ARQ and hybrid ARQ/FEC mechanism.
2. Along with the encoder and decoder, the additional non-neglected overhead components such as retransmission buffer and redundant information which may cause the increase of power consumption and network traffic load are considered when choosing an optimal error control mechanism for fault tolerance.
3. To suggest the proper choice for designer, the impact upon energy and performance of each error control schemes achieving the same reliability are compared simultaneously.

4.2 State-of-the-Art Reliable NoC

Error control schemes can be applied at either the packet or flit level. Most of the researches implement their error control scheme on flit level because of the relatively lower packet latency and less buffer needed at error control circuits [19]. DeMicheli et al. advocated using *ECC* to solve this problem at the data link layer for scalability [20]. Depending on the error recovery mechanism used, each input buffer may have to implement a decoder and at least each network interface must implement an encoder.

Error recovery can be done on a switch level or on an end-to-end level. When done on a switch level, after each hop, the input buffer has to perform a decoding procedure to check whether or not the neighboring buffer should be retransmitted. The end-to-end alternative is asking only the destination core to perform the decoding procedure and send an acknowledgement or an error packet back to the source. As concluded in [19], the end-to-end method would have a large impact on throughput especially if the average number of hops and the chance of errors were high. In those results, the switch-based retransmission method far outperformed the end-to-end based retransmission method and even the hybrid error detection/correction scheme (which corrects single bit error at the switch level and detects double bit errors with end-to-end recovery), as suggested in this chapter.

Since fault tolerance is likely to be a basic requirement for any large system, the design of the facilities for implementing it becomes an integral part of the system design process itself. This implies the cost and overheads associated with these facilities must be factored. Pullini et al. considered different error control protocols such as STALL/GO, T-Error, ACK/NACK and determined the overhead of providing such support when running in error-free environments [14]. Bertozzi et al. presented power versus performance results for point-to-point error control in an on-chip bus protocol based on AMBA schemes but not the real NoC architecture [18]. Ejlali et al. analyzed the impact of error control schemes on reliability, performance and energy objective when voltage swing varies [15]. However, all of these works did not consider the influence of network traffic load variance for different error control codes and neglected the power consumption and increased network traffic load affected by the retransmission buffer. In the later sections, we will provide different points of view and give a clear consideration of trade-off between the performance, energy, and reliability of an NoC architecture.

In this chapter, the codes used for comparison as candidates to instrument our NoC router with are simplistic linear block codes which have low overhead such as single error correction (*SEC*), double error detection (*DED*), single error correction and double error detection (*SECDED*), single parity (PAR) bit, and different lengths of cyclic-redundancy check (CRC) codes. Block codes can be written as (n,k) where n is the total number of bits in each codeword while k represents the number of useful information bits and n-k is the number of redundant bits for error control [21]. They are low implementation overhead codes which make them suitable for per router placement in NoC designs.

4.3 Fault Modeling

The reliability of a system is defined in International Telecommunication Union (ITU-T) recommendations E.800 as "The probability that an item can perform a required function under stated conditions for a given time interval". In this chapter, we will evaluate a hardware design based on reliability and a single event upset (SEU) fault model, which is capable of characterizing the effects observed in deep sub-micron designs and has been used to develop various fault tolerance mechanisms for NoC [22]. An SEU refers to one wire, independent of all other wires, being inverted for one clock cycle. This corresponds to a bit error in a transmitted flit on an NoC link. In Hegde and Shanbhag's work [4], the summation of uncorrelated noise sources in CMOS circuitry is formed as a Gaussian function affecting a particular victim wire. It is assumed that the gate output is in error when noise voltage V_N exceeds the gate decision threshold voltage which is about half of voltage swing V_{sw}. This model assumes V_N has a normal distribution with a variance of σ_N^2 with mean of 0. Therefore, the probability of a bit having an error $\in$ is shown in the following equation:

$$\epsilon = Q\left(\frac{V_{sw}}{2\sigma_N}\right),$$

where the function $Q(x)$ is a Gaussian pulse defined as:

$$Q(x) = \int_x^\infty \frac{1}{\sqrt{2\pi}} e^{-y^2/2} dy$$

According to the model defined, the probability with which a flit can error is equal to the probability of any of the bits having an error in a flit. Therefore, the flit error rate (*FER*) is defined as:

$$FER = 1 - (1 - \epsilon)^n,$$

where n is the length of a flit. By applying *ECC* function for fault tolerant, we define *REP* as the residual error probability as discussed in [18]. This is the probability of an undetected error which might cause failure in a system despite the application of an *ECC* (n, k). For example, if *SEC* is used, then residual flit error rate would describe the condition in which two bit errors happened on a single flit because all single bit errors would be corrected by *SEC*. In the case of *DED*, this error rate would describe when a triple bit error occurred since all double bit errors can be detected and retransmitted by the error recovery mechanism. Let m be the number of bits that this *ECC* can detect or correct, and ϵ_{ECC} be the probability of bit error. Then the *REP* of the *ECC* is:

$$REP_{ECC} = 1 - \sum_{i=0}^{m} \binom{n}{i} \epsilon_{ECC}^{i} (1 - \epsilon_{ECC})^{n-i}$$

By changing the summation according to the types of error that the *ECC* can detect or correct, this gives us the chance of a flit having an undetectable error for a given *ECC*.

As in an NoC, the *REP* value can be used to calculate the expected time to failure of a system. Given a clock frequency f, the flit injection rate i of the NoC, and the average number of hops h that a packet takes to reach its destination; we can calculate the mean time to failure (*MTTF*) for any given *ECC* as:

$$MTTF = \frac{1}{f \times i \times h \times REP_{ECC}}$$

Note that the *MTTF* value is specific with respect to the particular *ECC* used.

4.4 Energy Consumption in an NoC Architecture

The hardware router proposed in this chapter performs switch based retransmission for two reasons. First, we do not want an end-to-end retransmission implementation to counteract the positive effects on reducing average packet latency of our router decision flow. Furthermore, since the focus is on distributed router design for a very large or scalable NoC, the end-to-end retransmission becomes an unreasonable choice when the average number of hops increases.

Switch-to-switch level retransmission also comes in two forms: flit-based and packet-based. Admittedly multiple factors could influence the comparison of flit-based versus packet-based. In one series of tests performed [19], as long as packets contained four or more flits per packet then flit-based retransmission provides lower power results. This is restrictive and our tests do not utilize packets with less than four flits as this would reduce the capability of wormhole routing. Hence our hardware NoC router design uses switch-to-switch flit level retransmission.

4.4.1 Derivation of Energy Metrics

Evaluating a hardware design based on reliability depends on a fault model capable of characterizing the effects that are observed in deep sub-micron trends. In other fields such as wireless networks, average energy per useful bit (E_{pub}) has already been used to display performance tradeoffs [23]. For on-chip communication link [18], the amount of energy required to transfer one actual information bit from one router to the next is comprised of energy used to encode E_e, decode E_d, and transmit over the link E_t (including transmission of any retried flits) each scaled to one bit of information then summed up to find E_{pub}:

$$E_{pub} = E_e + E_d + E_t$$

To realize the energy usage of an NoC router, the components contributing to energy must be broken down and examined individually. The first is the encoder and decoder energy. As each packet is sent into the communication network from the network interface, it switches on the encoder logic. Next, energy is spent as the flit traverses the link. Then, because of switch-to-switch flit detection, it switches on the decoder logic per hop. When there is an error, if the decoder can correct then the flit can be sent to the input buffer. However, if the decoder cannot correct but detect the error, it must send a retransmit request back to the upstream buffer, which will switch on the retransmission logic. And when one flit is in error, a number of flits must be retransmitted because of delays specific to the NoC.

Therefore, for a typical NoC design, the equation of E_{pub} does not take into account the retransmission buffers impact on E_{pub}. In the next section, we will break down in more detail the energy consumed in an NoC and formulate a new representation in energy perspective.

4.4.2 Effect of Retransmission Buffer

The schematic of retransmission delay between switches as shown in Fig. 4.1 can be used to calculate the number of flits that must be retransmitted as an error occurs. The first cycle of delay incurred is to de-glitch the data at the upstream router before sending to the NoC link. This is important because NoCs may use different clock

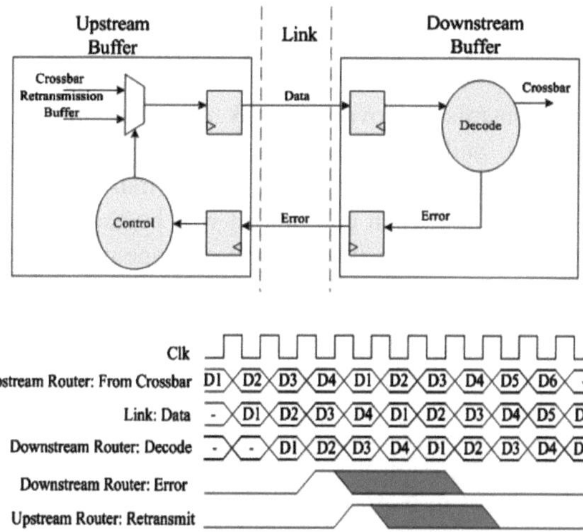

Fig. 4.1 **a** Schematic of retransmission delay over NoC link and **b** timing diagram of retransmission for NoC

domains or different clocking schemes to send data from router to router. The second cycle is registering the incoming data at the downstream router due to timing issues because of the length of the link. After the error is found, de-glitching the error at the downstream router causes another cycle of delay. And finally, at the upstream router due to the length of the link again, the error signal must be flopped.

This roundtrip causes a design which requires up to four flits to be re-sent when the transmitter responds to an error. The timing diagram is shown in Fig. 4.1b. In this figure, the signals correspond to different points along the schematic shown in Fig. 4.1a. The data are denoted as D1, D2, D3, ... etc. This diagram shows what happens when D1 is seen by the receiver as an error. Note: by the time the *Upstream Router: Retransmit Signal* is asserted, the upstream router has just sent out D4 on the previous cycle. The data provided to *Upstream Router: From Crossbar* will actually be switched on to feed from the retransmission buffer with data from four cycles ago, in other words, D1 will be re-sent. The main energy cost of the retransmission buffer is in the four sets of registers each a flit wide to latch data from four cycles ago. The simple implementation of the retransmission buffer causes the following flits from the erroneous flit to be resent, resulting in flits always being transmitted in order and no flit re-ordering on the receiver side. As shown in the shaded region, this design also has the benefit of the retransmission buffer disregarding errors for the following three cycles after an error. This can be good in designs where there are bursts of continuous errors on a single wire.

4.4.3 Re-Calculation of Energy per Useful Bit

So we append retransmission buffer energy E_r onto the previous equation for E_{pub} to derive the complete equation for E_{pub} :

$$E_{pub} = E_e + E_d + E_t + E_r,$$

where E_e is the average energy of the encoder, E_t is the average energy result of any transmission over the link over the link, E_d is the average energy of the decoder, and E_r is the average energy expended by the retransmission buffer module. All of these energies are normalized to per bit by taking into account the frequency that each component will be used, given a probability of bit error rate calculated from an input supply voltage and a fault model. All four components will be used during the retransmission of flits.

However, if this error control scheme has retransmission capabilities, it is reasonable that the retransmission buffer should be implemented with power saving considerations and be turned off when not in use. So we use E_{rl} to represent that the leakage power of E_r which is only consumed when there is no new data written into the retransmission buffer. Furthermore, we should consider the routing hops for each packet in NoC, this leads us to provide a more detailed version of E_{pub}:

$$E_{pub} = E_e + (E_d + E_t + E_r) \times h + ((E_d + E_t + E_{rl}) \times 4) \times p_r$$

In this equation, the energies are calculated in two parts, the energy of transmissions, and the energy of retransmissions multiplied by the probability density p_r. The p_r term only counts for the situations when the *ECC* can detect but cannot correct errors in a flit. From the p_r term calculation, we also modify two other things. First E_e is not counted since the router no longer needs to encode the flit again; it is sent straight from the retransmission buffer. Second, the calculation of the p_r needs to take into consideration the number of flits which are actually retransmitted per retransmission. Following the above discussion on the retransmission buffer, a factor of four is used since four flits are retransmitted per error in our design. However, different retransmission designs could lead to different factors. And the factor h represents the average number of hops that a packet takes to reach its destination.

4.5 Experimental Results

A complete hardware router design was implemented in HDL and synthesized using Synopsys Design Compiler. PrimePower was used to measure the energy of each component with a sampling interval of 0.1 ns under TSMC 0.13 μm process.

4.5.1 Experiments Setup

The standard deviation of the noise voltage was set as 100 mV. The codes used for comparison as candidates to instrument our NoC router with were simple linear block codes. For reference, an NoC design without encoding or decoding abilities was also used in the test results. The physical layer of our network comprised of 8×8 nodes each with 5 input buffers and 5 output ports. Each buffer was 8 flits long while each packet had a constant of 8 flits and each flit had 32 bits. Each output port contained four sets of registers as the retransmission buffer to latch data from four cycles ago. We assumed a packet injection rate of 0.01 packets per cycle for each node. The packet injection rate given in our design represents a packet injection rate that heavily utilizes the NoC yet stays under the saturation point.

4.5.2 Error Control Codes used in Experiments

In the following experimental results, six kinds of error control schemes were implemented with a baseline NoC router and the useful information bit was set as 32 for comparison.

The unencoded design (*UNENC*) and the single error correction (*SEC*) coding are two types of coding which do not require a retransmission buffer in the router design. The retransmission buffer is not needed when the set of correctable transmissions contains the set of detectable transmissions. The *SEC* coding is a basic implementation of a (38, 32) Hamming code. The Hamming code has a distance of three between each codeword, therefore it can be used as a single error correcting code. This implies the use of a correction circuit to output the true codeword has the least distance from the received codeword.

Alternately, the (38, 32) Hamming code can be used as *DED* to detect up to double the amount of errors. The decoding circuitry is similar to the *SEC* code but when a received codeword does not match with a true codeword, no attempt is made to correct the codeword and instead a retransmit signal is issued by the receiver. Thus a received codeword which has one or two bit errors can always be detected since this code has a distance of three. Since this code detects up to two bit errors and does not correct one bit nor two bit errors, its usage includes a retransmission buffer design.

A (39, 32) code can be constructed by adding an extra check bit to include the even parity of all the other bits. This creates a *SECDED* code with a minimum difference of 4 which has the ability to detect and correct a single error and at the same time detect but not correct a double error. If the received codeword is one bit off from a real codeword, it can be matched to that real codeword and the retransmission buffer can remain inactive. However, if it lies within a distance of two between two real code words, then a retransmit signal to the retransmission buffer has to be issued as the decoding circuit cannot identify which real codeword to choose.

To compare an *ECC* method with a small amount of check bits, we also designed a (33, 32) *PAR* code which simply adds a check bit to make the sum of the transmitted codeword bits even. It can detect single bit errors and require the use of a retransmission buffer whenever the error occurs.

Table 4.1 Characteristics of error control codes

Error control codes	Area (μm2)		Power (μW)		Delay (ns)	
	ENC	DEC	ENC	DEC	ENC	DEC
CRC4	629.74	563.54	282.54	300.55	0.26	0.27
CRC8	818.14	891.13	376.34	369.43	0.21	0.28
DED	719.70	1057.48	470.14	717.53	1.27	1.32
PAR	419.26	526.19	285.11	368.93	0.89	0.97
SEC	719.70	1933.34	470.14	1136.80	1.27	1.93
SECDED	796.08	2636.07	512.92	1914.40	0.87	1.90

Finally, we considered two *CRC* codes. *CRC* code words can be cyclically shifted to produce other valid code words. They are widely used in computer networks and provide burst-error detection capabilities. The *CRC* code is completely specified by a generator polynomial which information bits are multiplied with to create the code words. The degree of the polynomial used is the number of check bits added to the code. The *CRC* codes that we considered here use the following generator polynomials:

$$G(x) = x^4 + 1$$

$$G(x) = \mathrm{x}^8 + 1$$

to represent a *CRC* code of degree 4 (*CRC*4) and a *CRC* code of degree 8 (*CRC*8), respectively. The *CRC*4 code is a (36, 32) code that can detect burst errors of less than or equal to 4, while the *CRC*8 code is a (40, 32) code that can detect burst errors of less than or equal to 8.

Note that two of the codes, *UNENC* and *SEC* do not have retransmission buffers in their designs, which represent a substantial saving in area. All the other types of codes that use error detection need retransmission buffers on their NoC routers.

The area, power, and delay information of both the encoder (ENC) and decoder (DEC) parts for each of the hardware implemented error control codes are illustrated in Table 4.1. We can see that the area and power in each of the encoders is not quite different, but the decoders with error correcting function (such as *SEC* and *SECDED*) have much more overhead. Also, hamming codes have more overhead in timing delay than *CRC* codes, especially those with error correcting functions.

4.5.3 Results Analysis

Table 4.2 shows the energy consumption of different portions of the router when the *MTTF* was set as one year and the wire load lengths of the links used between routers were modeled from 500 μm to 2,000 μm as the useful information data bit *k* are all set as 32. We can see that the energy consumed in retransmission buffers

Table 4.2 Composition of energy expenditures

Error control codes	Energy (J)						
	Encoder	Decoder	Retransmission	Link (500 μm)	Link (1,000 μm)	Link (1,500 μm)	Link (2,000 μm)
CRC4	1.4411E-14	1.6879E-14	6.0413E-13	1.7051E-13	2.7859E-13	4.2153E-13	5.7085E-13
CRC8	1.8769E-14	1.7820E-14	6.7125E-13	1.8946E-13	3.0955E-13	4.6836E-13	6.3427E-13
DED	1.3219E-14	2.4604E-14	6.3769E-13	1.7998E-13	2.9407E-13	4.4495E-13	6.0256E-13
PAR	5.2313E-15	7.5450E-15	5.3700E-13	1.5630E-13	2.5538E-13	3.8640E-13	5.2328E-13
SEC	1.3219E-14	6.3825E-14	N/A	1.7998E-13	2.9407E-13	4.4495E-13	6.0256E-13
SECDED	1.6226E-14	9.8963E-14	6.5447E-13	1.8472E-13	3.0181E-13	4.5666E-13	6.1842E-13

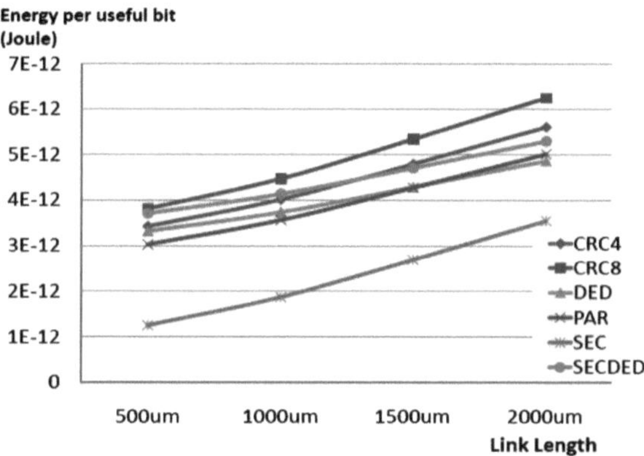

Fig. 4.2 E_{pub} versus link length when *MTTF* is set as one year

and links were higher than that of the encoder and decoder by about one order of magnitude. And the energy consumed in retransmission buffers can even be two to three times higher than in the links. Hence the retransmission buffers really play an important role in power dissipation.

To illustrate the effect of power dissipation in retransmission buffer more clearly. We can see in Fig. 4.2 that *SEC* takes about only 25–30% energy consumption than other error control code schemes while link length is set as 500 μm. Furthermore, as the link length becomes longer, the E_{pub} for *SEC* is still much better than the others since there is no energy expenditure from the retransmission buffer.

For the other error control mechanisms that need retransmission buffers, the redundant bits (*n–k*), used for error control is the main reason for energy expenditure when the link length is not very long. In other words, the more redundant bits, the more energy it will need to consume. As the link length becomes longer, the effect of power on link will become more important. And we can find that the energy expenditure for those error control schemes that can detect less error

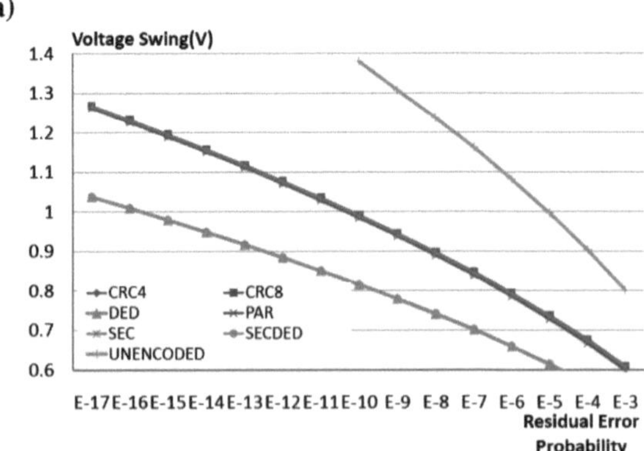

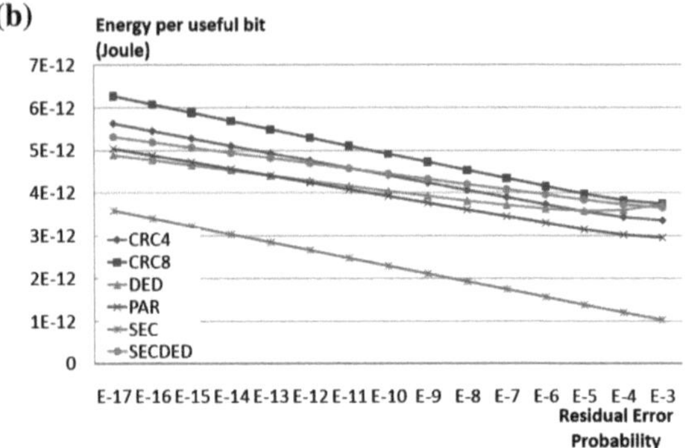

Fig. 4.3 **a** Voltage swing versus *REP* and **b** E_{pub} versus *REP* while the link length set as 2,000 μm

increase faster than others. Since the *MTTF* for each *ECC*s were all set as one year, the different *ECC* capability means different needs of voltage swing on inter-router interconnection wires. In other words, the voltage swing on link for data transmit will affect the value of E_{pub} more as the link length increases. Also, while the link length is getting longer, the impact of retransmission buffer is getting less and the E_{pub} of *SEC* is getting closer to others.

To see how the error control capability of each *ECC* affects the voltage swing needed to transmit data, Fig. 4.3a shows the voltage swing versus residual error probability (*REP*) view for reference.

REP defines how often a flit would be undetected and uncorrected in an NoC. This graph can help get the minimum voltage that each *ECC* should use to meet the same reliability constraints. We can observe that the number of errors that an *ECC* promises is the main impact factor to voltage swings. Hence the error control schemes can be divided into three groups. Under the same reliability constraint, the *UNENCODED* scheme needed more energy to increase signal to noise ratio. On the other hand, the error control schemes which have higher error protection capability such as *DED* and *SECDED* needed less energy than others.

In addition to voltage needs for interconnection links, Fig. 4.3b illustrates the E_{pub} needs under different *REP* for different error control schemes for comparison.

Table 4.3 Operating conditions for three *MTTF* intervals

Error control codes	Codeword size	MTTF = 1 day		MTTF = 1 month		MTTF = 1 year	
		pr	*Vsw*	*pr*	*Vsw*	*pr*	*Vsw*
CRC4	36	4.24 E-07	1.164	7.6 4E-08	1.22	2.14 E-08	1.26
CRC8	40	4.18 E-07	1.168	7.4 9E-08	1.224	2.09 E-08	1.264
DED	38	1.33 E-04	0.956	4.1 4E-05	1	1.88 E-05	1.032
PAR	33	4.38 E-07	1.16	7.45 E-08	1.216	2.23 E-08	1.256
SEC	38	0	1.164	0	1.22	0	1.26
SECDED	39	2.28 E-09	0	1.9 8E-10	1.002	4.06 E-11	1.034
UNENC	32	0	1.62	0		0	

Under the same *REP* constraint, we can still find that those *ECC*s that needs retransmission buffers has a higher E_{pub} expenditure than *SEC*. And, as the reliability constraint is getting lower, we can find the slope change of *DED* curve. This is because the value of *pr* will become larger as the *REP* is getting larger and cause the effect of the second part of the last equation of E_{pub} to be more important.

The *MTTF* can be calculated by evaluating the number of wires in an NoC and the probability of faults each wire may have. In addition to running experiments at 1.08 V, we need to standardize the energies to represent the same *MTTF*. Since the more errors a code can handle the longer *MTTF* it will have, a code that can handle more errors can transfer data at lower voltage while sustaining the same *MTTF*. Table 4.3 shows the voltage swing needs and the respective retransmission probability *pr* for each error control scheme under the three *MTTF* intervals.

In Fig. 4.4, we evaluate the *MTTF* for different *ECC*s at the three intervals. We can see that *SEC* which has no retransmission buffer performs better in both Fig. 4.4a for a 500 μm link, and Fig. 4.4b for a 2,000 μm link, under all *MTTF* constraints. Furthermore, Fig. 4.4 also shows the same result as Fig. 4.2 that redundant bits contribute more to E_{pub} as wire length is not as long. However, as the wire length gets longer, the effect of voltage swings will become important to E_{pub}.

To get a fair comparison between different error control schemes under the same hardware resources such as the same bandwidth of buffer and inter-router transmission links, average energies of the respective *ECC*s were found and plotted against the amount of effective bandwidth that an NoC can provide. In the previous experiment, we used extra links for redundant bits of error control codes. However, if the bandwidth of an NoC has been regularized, the whole codeword including information bit and redundant bit for error control should be put into each flit of the packet and transmitted. In other words, a series of data will need longer packets, or more flits, if the *ECC* used has more redundant bits.

Figure 4.5a shows the Throughput Efficiency for each *ECC*, which can be calculated by:

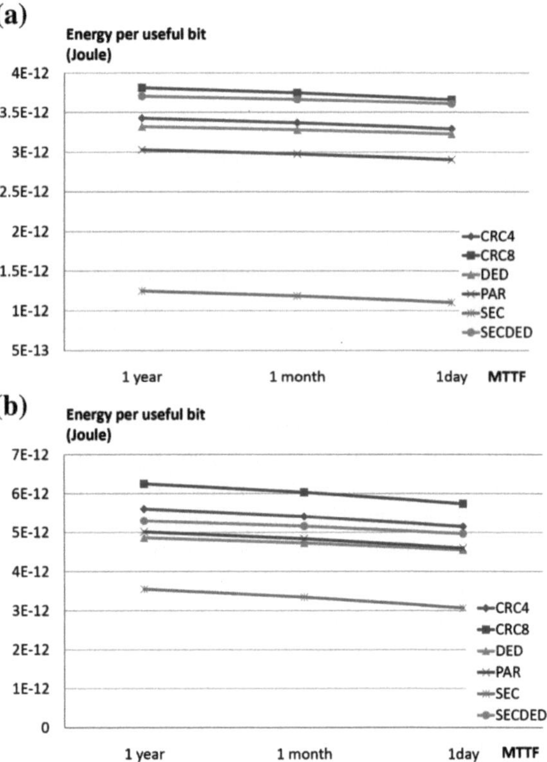

Fig. 4.4 E_{pub} versus *MTTF* where the link length is modeled as **a** 500 μm and **b** 2,000 μm

$$\text{Throughput} = \frac{K}{N} \times \left(\frac{1}{1 + p_r}\right),$$

where K is the useful information bits that do not include the redundant bits for error control sent over the link and N is the bandwidth allocated between routers. This is important because even if an *ECC* can achieve a very low E_{pub}, it is of no use if it degrades the throughput of the link by so much that the link must be widened several times. Also, more flits induce higher data injection rate to the NoC and cause the performance of average data transmission latency to decrease exponentially. Therefore, the redundant bits for error control also should be considered while designing a reliable NoC.

In this situation, *SEC* still performs well in the view of E_{pub} versus bandwidth as illustrated in Fig. 4.5b. As for those error control schemes with retransmission buffers, we can see that the number of redundant bits dominate the E_{pub} value as the bandwidth is narrow because the redundant bit will cause the packet length to become longer to transmit. However, as the bandwidth expands under the same reliability constraint, E_{pub} value for each error control scheme will saturate.

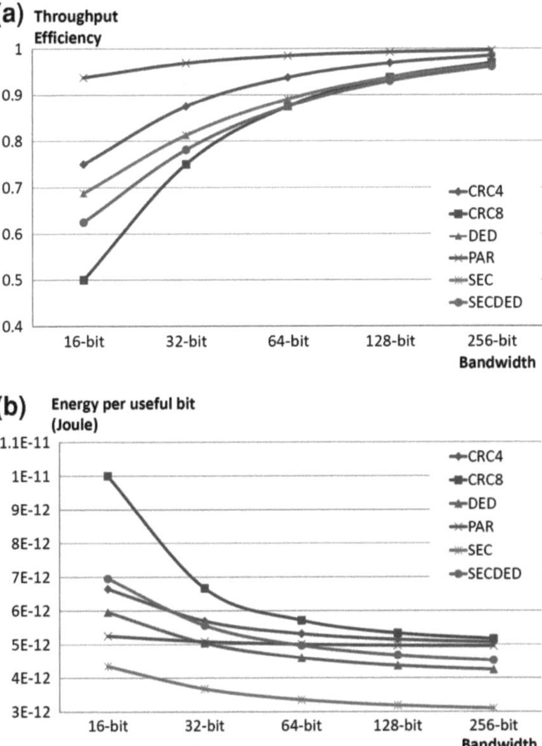

Fig. 4.5 Comparisons of each *ECC* in both **a** throughput and **b** energy dimensions versus bandwidth while th *MTTF* is set as one year and link length is set as 2,000 μm

4.6 Remarks

Retransmission has a substantial impact on energy. Our hypothesis that retransmission is not as favorable as previously thought can be proved by the experimental results. While the retransmission buffer was put into consideration, the energy consumed in encoder and decoder becomes less important. This is because the retransmission buffer and link length are the cause of having a higher energy expenditure compared to the other factors. All aspects of views we provided show that the retransmission buffer is always the main reason for energy dissipation. Therefore, simple schemes without any retransmission technique such as *SEC* should be considered for use in NoC routers to achieve circuit reliability. Considering both the power consumption and the complex implementation of a retransmission protocol, the choice will depend on the wire length used and the *MTTF* required in a system within a reasonable bandwidth constraint. While the bandwidth constraint is very narrow, multiple-bit error detection schemes that need a retransmission mechanism can be considered to improve the network traffic tolerance. Under the same *MTTF*, the longer the link length between routers, the more voltage swing (depending on the error control capability of *ECC*s) impacts

E_{pub}. Therefore, the *ECC* that can handle more errors is more suitable for longer link lengths while the *ECC* that needs less redundant bits is more suitable for shorter link lengths.

References

1. G. Martin, "Design Methodologies for System Level IP", in Proceedings of the Conference on Design, Automation and Test in Europe, pp. 286–289, March 1998
2. J. Nurmi, H. Tenhunen, J. Isoaho, and A. Jantsch, *Interconnect-Centric Design for Advanced SOC and NOC*, Springer, 2004
3. R. Ho, K. W. Mai, and M. A. Horowitz, "The Future of Wires", Proceedings of the IEEE, vol. 89, no. 4, pp. 490–504, April 2001
4. R. Hegde and N. R. Shanbhag, "Toward Achieving Energy Efficiency in Presence of Deep Submicron Noise," IEEE Transactions on Very Large Scale Integration Systems, vol. 8, no. 4, pp. 379–391, August 2000
5. C. Constantinescu, "Trends and Challenges in VLSI Circuit Reliability", IEEE Micro, vol. 23, no. 4, pp. 14–19, July 2003
6. N. Cohen, T. S. Sriram, N. Leland, S. Butler, and R. Flatley, "Soft Error Considerations for Deep-Submicron CMOS Circuit Applications", in Proceedings of the International Electron Devices Meeting Technical Digest, pp. 315–318, December 1999
7. P. Shivakumar, M. Kistler, S. Keckler, D. Burger, and L. Alvisi, "Modeling the Effect of Technology Trends on the Soft Error Rate of Combinational Logic", in Proceeding of the Dependable Systems and Networks, pp. 389–398, June 2002
8. D. Sylvester, "A Global Wiring Paradigm for Deep Submicron Design", IEEE Transactions on Computer Aided Design of Integrated Circuit and Systems, vol. 19, no. 2, pp. 242–252, February 2000
9. R. Marculescu, U. Y. Ogras, L. S. Peh, N. E. Jerger, and Y. Hoskote, "Outstanding Research Problems in NoC Design: System, Microarchitecture, and Circuit Perspectives", IEEE Transactions on Computer-Aided Design of Integrated Circuits and Systems, vol. 28, no. 1, pp. 3–21, January 2009
10. P. Vellanki, N. Banerjee, and K. S. Chatha, "Quality-of-Service and Error Control Techniques for Mesh-Based Network-on-Chip Architectures", ACM Very Large Scale Integration Journal, vol. 38, no. 3, pp. 353–382, January 2005
11. T. Bjerregaard and S. Mahadevan, "A Survey of Research and Practices of Network-on-Chip", ACM Computing Surveys, vol. 38, no. 1, pp. 1–51, March 2006
12. J. Duato, S. Yalamanchili, and L. Ni, Interconnection Networks: An Engineering Approach, Morgan Kaufmann, 2002
13. A. Jantsch and H. Tenhunen, Networks on Chip, Kluwer Academic, 2003
14. A. Pullini, F. Angiolini, D. Bertozzi, and L. Benini, "Fault Tolerance Overhead in Network-on-Chip Flow Control Schemes", in Proceeding of the Symposium on Integrated Circuits and Systems Design, pp. 224–229, September 2005
15. A. Ejlali, B.M. Al-Hashimi, P. Rosinger, and S. G. Miremadi, "Joint Consideration of Fault-Tolerance, Energy-Efficiency and Performance in On-Chip Networks", in Proceedings of the Conference on Design, Automation and Test in Europe, pp. 1647–1652, April 2007
16. F. Worm, P. Ienne, P. Thiran, and G. DeMicheli, "A Robust Self-Calibrating Transmission Scheme for On-Chip Networks", IEEE Transactions on Very Large Scale Integration Systems, vol. 13, no. 1, pp. 126–139, January 2005
17. C. Svensson, "Optimum Voltage Swing on On-Chip and Off-Chip Interconnect", IEEE Journal of Solid-State Circuits, vol. 36, no. 7, pp. 1108–1112, July 2001

18. D. Bertozzi, L. Benini, and G. DeMicheli, "Error Control Schemes for On-Chip Communication Links: the Energy-Reliability Tradeoff", IEEE Transactions on Computer-Aided Design of Integrated Circuits and Systems, vol. 24, no. 6, pp. 818–831, June 2005
19. S. Murali, T. Theocharides, N. Vijaykrishnan, M. J. Irwin, L. Benini, and G. DeMicheli, "Analysis of Error Recovery Schemes for Networks on Chips", IEEE Design & Test of Computers, vol. 22, no. 5, pp. 434–442, September 2005
20. G. DeMicheli and L. Benini, Networks on Chips: Technology and Tools, Morgan Kaufmann, 2006
21. S. Lin and D. J. Costello, Error Control Coding, Prentice-Hall, 1983
22. J. Hu and R. Marculescu, "Energy-Aware Communication and Task Scheduling for Network-on-Chip Architectures under Real-Time Constraints", in Proceedings of the Design, Automation and Test in Europe Conference and Exhibition, pp. 234–239, February 2004
23. E. Shih, "Physical Layer Driven Protocol and Algorithm Design for Energy-Efficient Wireless Sensor Networks", in Proceedings of the International Conference on Mobile Computing and Networking, pp. 272–287, July 2001

Chapter 5
Energy-Aware Task Scheduling for Noc-Based DVS System

For real time applications, time slacks of a preliminary task schedule may be exploited to conserve energy. This can be accomplished by leveraging the dynamic voltage scaling (DVS) technique to slow down clock frequency of certain cores as long as the deadline is met. In this chapter, the task of fine-tuning an existing task assignment and schedule and using DVS to lower the overall energy consumption is formulated as a graph-theoretic maximum weight clique (MWC) problem. An efficient heuristic algorithm is proposed to systematically solve this problem. A unique feature of our approach is concurrently applying DVS to slow down the execution of multiple tasks to achieve better energy savings. Extensive simulations are performed to compare this proposed algorithm against leading energy-aware task scheduling algorithm and DVS algorithm. Our algorithm exhibits 22% more energy savings than the Energy Aware Scheduling (*EAS*) algorithm. As for energy saving in DVS process, our MWC-based method provides a 97% saving improvement over the PV-DVS algorithm.

5.1 Problem Formulation

Our goal in this chapter is to develop a new algorithm for energy-aware scheduling considering both communication and computation for NoC architectures while using DVS technique to minimize the energy consumption. Given an application task graph and an NoC architecture, we want to find (1) an energy-aware scheduling of tasks to PEs, such that all the hard deadline constraints are met, and (2) an power optimization algorithm to utilize the slack time for energy saving. Compared to existing works, our work has made several tangible contributions:

1. A Novel Energy-Aware Scheduling (NEAS) algorithm including task prioritization and task assignment steps for energy optimization while taking into account the NoC architecture.

S.-J. Chen et al., *Reconfigurable Networks-on-Chip*,
DOI: 10.1007/978-1-4419-9341-0_5, © Springer Science+Business Media, LLC 2012

2. Integrating DVS into the task scheduling algorithm; and adjusting the scheduling results during power optimization process iteratively to increase slack utilization.
3. A Maximum Weight Clique based DVS (MWC-DVS) problem formulation and an MWC-DVS heuristic algorithm for solving the problem.
4. An efficient re-scheduling technique to adjust the scheduling along with power optimization process.

5.1.1 Application and Architecture Specification

To deal with the scheduling of an application on a specialized NoC architecture, we will first introduce some definitions that will be used in this chapter.

Definition 1 A *task graph* $G = G(T, E)$, is a directed acyclic graph that consists of a set of vertices T and directed edges E. Each vertex $\tau_i \in T$ represents a computational task of an application that needs to be executed on a processor core. Each directed edge $e_{i,j} \in E$ represents an inter-task data dependency where data will need to be transferred from the ith task to the jth task via an NoC fabric.

Two tasks can be executed concurrently at different processor cores if there is no directed path in the task graph G linking one to the other. A node τ_i in G will be assigned to a processor core for execution. The execution time and energy consumed of executing τ_i in a processor core at particular clock frequency are assumed to be known. A *deadline* $dl(\tau_i)$ counted from the starting time of the entire application will also be set in advance. The execution of τ_i must be completed before $dl(\tau_i)$ to ensure correctness of the result. Each directed arc $e_{i,j} \in E$ dictates that the task v_j must not start before v_i is finished. Each $e_{i,j}$ has an associated label $vol(e_{i,j})$, reflecting the quantity of information to be forwarded from task τ_i to τ_j.

Definition 2 The NoC architecture model is generally specified as a directed graph $A(P, CL)$ to represent the processing elements (PE) and switches that are connected by a specific network topology represented by communication links (CL) in the platform. Each vertex $p_i \in P$ denotes a processing element that is annotated with relevant information with respect to the type of a processor. For power management facility, each p_i could be a state-of-the-art voltage scalable component which has capability to dynamically switch among a set of available levels of supply voltages, $v_{sw_j}^{p_i} \in V_{sw}^{p_i}$ during different time intervals. Power management techniques could be employed to totally or partially shut down a PE or CL. Each directed arc $l_{i,j} \in CL$ represents a data transmission link from p_i to p_j, which is also associated with information such as data bandwidth, $bw(l_{i,j})$, and energy consumption, $e(l_{i,j})$, on one bit of data transmitting from p_i to p_j.

5.1.2 Generalized Energy-Aware Task Scheduling Problem

Scheduling is the order in time of computation and communication actions on their assigned resources, which assures the mutual exclusion of any execution on the same resource at any moment. Recently, much research has been focused on energy-aware scheduling techniques for real-time system, instead of only maximizing system performance. However, to cope with many-core systems, complex communication network should also be considered during task scheduling. Since both communication transactions and task execution need to be considered, we can describe the energy-aware scheduling problem for an NoC architecture under real-time constraints as follows:

$$\min\left\{\sum_{\forall\tau_i\in T} E_{M(\tau_i)} + \sum_{\forall e_{i,j}} vol(e_{i,j}) \times E_{RA\left(M(\tau_i),M(\tau_j)\right)}\right\}$$

such that

- All the deadline constraints $dl(\tau_i \in T)$ are satisfied.
- All the task dependences are compliant.
- All the computations and communications assure mutual exclusion policy.

$E_{M(\tau_i)}$ represents the computation energy in mapping task τ_i on PE $M(\tau_i)$. Given the start time $\left(T_s^{M(\tau_i)}\right)$ and end time $\left(T_e^{M(\tau_i)}\right)$ constraints of running τ_i on $M(\tau_i)$, the computation energy for voltage scalable processing element can be further illustrated as:

$$E_{M(\tau_i)} = \sum\left(v_{sw_1}^{M(\tau_i)}(\Delta t_1), v_{sw_2}^{M(\tau_i)}(\Delta t_2), \ldots, v_{sw_j}^{M(\tau_i)}(\Delta t_j)\right),$$

where $v_{sw_j}^{M(\tau_i)}(\Delta t_j)$ represents running PE, $M(\tau_i)$ at operating voltage $v_{sw_j}^{M(\tau_i)}$ for Δt_j duration, and $(T_e^{M(\tau_i)} - T_s^{M(\tau_i)}) = \sum_{k=1}^{j} \Delta t_k$ is the total execution time of task τ_i on $M(\tau_i)$. For simplification, we neglect the voltage scaling overhead in terms of power and time in this work.

Moreover, let $E_{RA\left(p_i,p_j\right)}$ represent the energy consumption for one bit transmission from p_i to p_j under a specific routing path allocation algorithm. Given a different task scheduling result, its inter-task routing path and traffic congestion also vary; thus, we divide the communication energy spent for one bit transmission into E_s, E_L and E_B to respectively represent the energy consumed on the switch, on the link, and waiting, which can be calculated as:

$$E_{RA\left(p_i,p_j\right)} = n_{hops} \times E_s + \left(n_{hops} - 1\right) \times E_L + n_{cong} \times E_B,$$

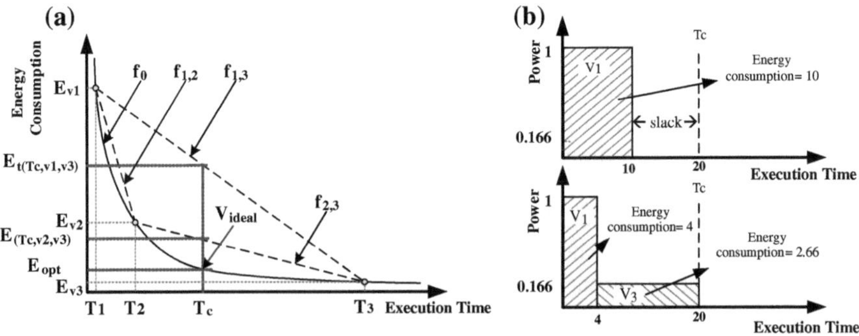

Fig. 5.1 Energy consumption versus time constraint

where n_{hops} is the number of hops from p_i to p_j decided by the routing algorithm, and n_{cong} is the number of congestion cycles that the data need to wait in the buffer during transmission after the scheduled communication.

5.1.3 Dynamic Voltage Scaling

The main idea behind DVS is to scale the supply voltage V_{dd} and operating frequency f dynamically according to the real-time performance requirements of the application. Since the dynamic power consumption is the dominant source of power dissipation in a digital CMOS circuit, first we will describe the relation of dynamic power and operational voltage as:

$$P_{dyn} = C_L \times N_{sw} \times V_{dd}^2 \times f$$

$$f = k \times ((V_{dd} - V_t)^\alpha / V_{dd}),$$

where C_L denotes the load capacitance of the circuit, N_{sw} is the number of switching activity, V_{dd} is the supply voltage, α is a parameter between 1.0 and 2.0 depending on technology process, and k is the circuit dependent constant. Thus, we can find that P_{dyn} depends quadratically on V_{dd}. And reducing V_{dd} is the most effective way for power reduction but it may lower the operating frequency and increase the computation delay.

Ishihara et al. [1] has proved that if a processor can be operated using discretely variable voltages, the voltage scheduling with at most two voltages, which are immediate neighbors to the voltage V_{ideal}, minimizes the energy consumption under any time constraints. As illustrated in Fig. 5.1a, a clear view of the relation between energy consumption and timing constraint for voltage scalable processor is shown.

We can find that, under timing constraint T_c, the optimized energy consumption E_{opt} can only be obtained by running at ideal operating voltage V_{ideal}. If the operating voltage of a processor can only be dynamically scaled under three voltages, V_1, V_2, and V_3, the minimum energy consumption obtained under timing constraint T_c is $E_{(T_c,v_2,v_3)}$, which denotes the energy consumption running at the optimized voltage scaling of V_2 and V_3. In other words, DVS technique essentially varies with the timing constraints T_c and we need to find the energy equation $f_{i,j}(T_c)$ which is closest to ideal energy consumption $f_0(T_c)$. As illustrated in Fig. 5.1b, assume there are only two voltage levels v_1 and v_3 for use, the total energy consumption is 10 units of energy under voltage v_1, even if the power supply is turned off after finishing the program. Given a time constraint of T_c, the voltage scheduling with v_1 and v_3 which fits the execution time with the given T_c reduces the energy consumption from 10 units of energy to 6.66 units of energy.

Therefore, instead of finding optimum energy under timing constraint T_c on curve f_0 such as the conventional algorithm does, we can find the energy saving rate $\left(S_{i,j}^{M(\tau_k)}\right)$ for a given task τ_k on $M(\tau_k)$ between discrete voltages v_i and v_j by the following energy consumption $\left(E_{i,j}^{M(\tau_k)}\right)$ and execution time $\left(T_{i,j}^{M(\tau_k)}\right)$ equations:

$$E_{i,j}^{M(\tau_k)}(x) = c_{M(\tau_k)}^{E} \times (v_i^2 \times x + v_j^2 \times (EC_{M(\tau_k)} - x))$$

$$T_{i,j}^{M(\tau_k)}(x) = c_{M(\tau_k)}^{T} \times \left(\frac{x \times v_i}{(v_i - v_t)^{\alpha}} + \frac{\left(EC_{M(\tau_k)} - x\right) \times v_j}{(v_j - v_t)^{\alpha}}\right)$$

$$S_{i,j}^{M(\tau_k)} = \frac{\Delta E_{i,j}^{M(\tau_k)}}{\Delta T_{i,j}^{M(\tau_k)}} = \frac{E_{i,j}^{M(\tau_k)}\left(EC_{M(\tau_k)}\right) - E_{i,j}^{M(\tau_k)}(0)}{T_{i,j}^{M(\tau_k)}\left(EC_{M(\tau_k)}\right) - T_{i,j}^{M(\tau_k)}(0)},$$

where $EC_{M(\tau_k)}$ stands for the total execution cycle of a given task τ_k on $M(\tau_k)$ and x represents the execution cycles on voltage v_i. Therefore, allocating slacks to a processor which has higher value of $S_{i,j}^{M(\tau_k)}$ gets more energy saving.

5.2 Motivational Example

Motivational example is illustrated in this section to more clearly explain our contribution. Each node in Fig. 5.2 represents a task in the given application and a directed arrow indicates data dependency. To make this example easier to understand, we will use four homogeneous fully connected PEs (P1, P2, P3, and P4) as our simulation architecture. Assume there are two voltage levels (v_{max} and v_{min}) for voltage scaling, Fig. 5.2b shows the execution time (exe_t) for each task running at v_{max}, and energy saving rate, $S_{v_{max},v_{min}}^{M(\tau_k)}$ for reference. Communication delay and

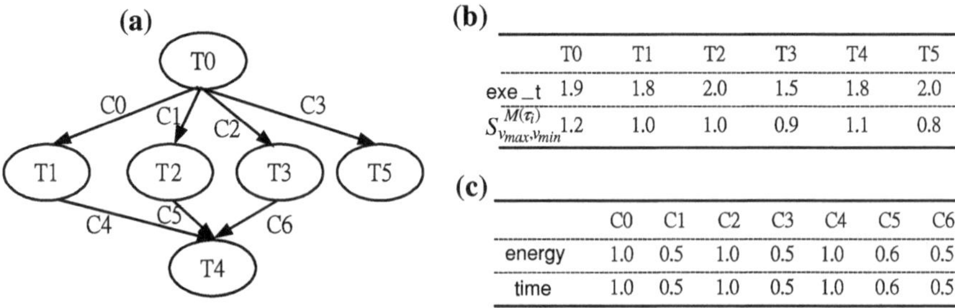

(b)

	T0	T1	T2	T3	T4	T5
exe_t	1.9	1.8	2.0	1.5	1.8	2.0
$S_{v_{max},v_{min}}^{M(\tau_i)}$	1.2	1.0	1.0	0.9	1.1	0.8

(c)

	C0	C1	C2	C3	C4	C5	C6
energy	1.0	0.5	1.0	0.5	1.0	0.6	0.5
time	1.0	0.5	1.0	0.5	1.0	0.6	0.5

Fig. 5.2 Task graph of the motivational example

energy for each arc are shown in Fig. 5.2c while these two dependent tasks are mapped to different PEs.

In order to reduce the communication energy, most of the conventional task scheduling algorithms will map tasks on the same PE as long as the deadline restriction is not violated as shown in Fig. 5.3a. However, for a dynamic voltage scalable system, if we can change the mapping of T2 to P1 as illustrated in Fig. 5.3b, there will be more slack left for energy saving and we can probably get 4.8 units of energy saving as illustrated in Fig. 5.3c. Hence, if we can estimate the potential of energy saving with a DVS process; better energy results may be obtained by trading off with communication penalty.

In observation of the slack distribution in Fig. 5.3d, we can find that T0 has highest energy saving rate $S_{v_{max},v_{min}}^{M(\text{T0})} = 1.2$, but extending T0 will postpone the two end nodes of T4 and T5. That is, extending one unit-time at T0 costs two unit-times of slacks. On the other hand, extending one unit-time at T1, T2 and T5 only costs two unit-times of slacks as illustrated in Fig. 5.3g which is much more efficient than the conventional methodology for energy saving. Therefore, we will propose a new algorithm on allocating slacks for energy saving using a DVS process while considering the dependencies in a task graph to provide efficient energy saving in a greedy manner in Sect. 5.3.

Furthermore, since the iterative DVS process will cause the change of communication traffic after the insertion point of the slack. It is reasonable to re-schedule the task after the slack insertion point for better energy dissipation. However, most of the previous work did not consider the effect of slack insertion and remain the same task execution orders at all PEs and may result in some unpredictable penalty. Here, we re-consider the example as shown in Fig. 5.3 and illustrate the benefit of re-scheduling tasks in Fig. 5.4.

Figure 5.4a shows the slack allocation after the first iteration of DVS process. If we keep the original task execution order in each PE to do the next iteration of DVS process, then we can get 4.8 units of energy saving as shown in Fig. 5.3i. However, if we can re-consider the task execution time after the insertion point of slack, T3 may have chance to be re-scheduled to P4 since the potential energy saving will be better as shown in Fig. 5.4b. Thus, total energy saving can be obtained up to 7.6 units of energy as shown in Fig. 5.4c after re-scheduling.

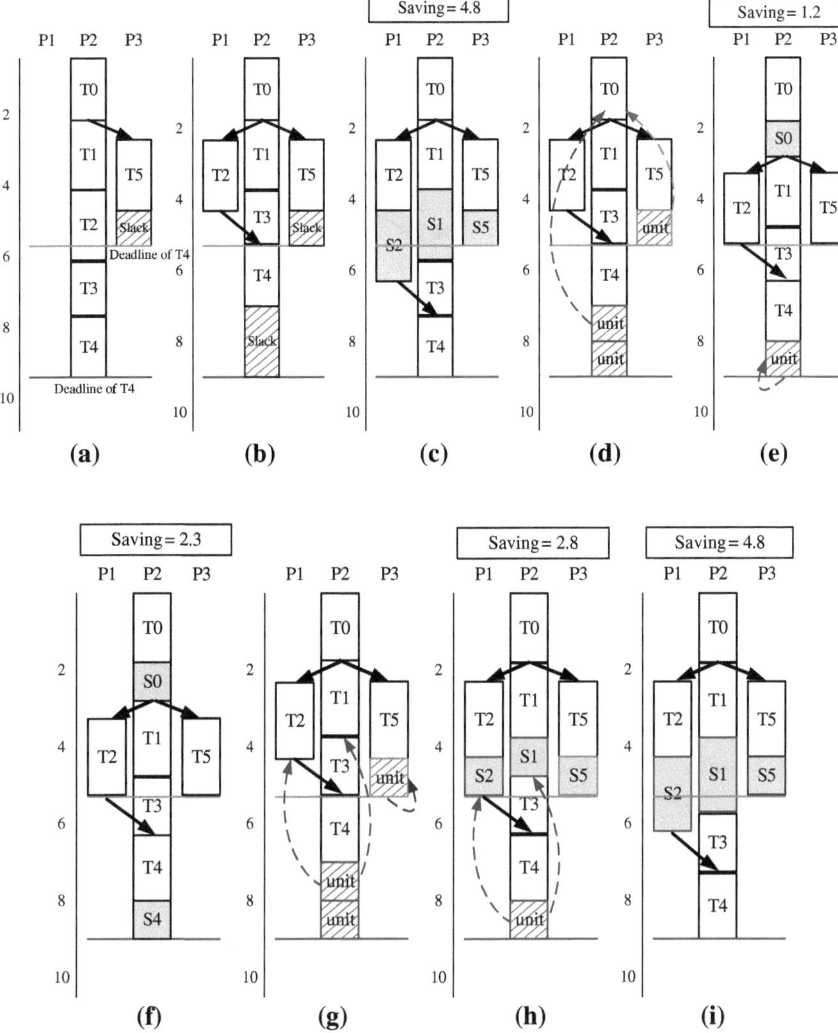

Fig. 5.3 Motivational example of our energy-aware task scheduling algorithm

5.3 Proposed Algorithmic Solution

In this work, the energy-aware task scheduling problem includes assigning the tasks of an application to suitable processors and ordering task executions on each resource. For power management at early stage, we integrate DVS consideration into our algorithm as illustrated in Fig. 5.5. An ordered list of tasks is constructed by assigning a priority to each task first. Then, tasks are selected in the order of their priorities and each selected task is scheduled to a processor which minimizes a

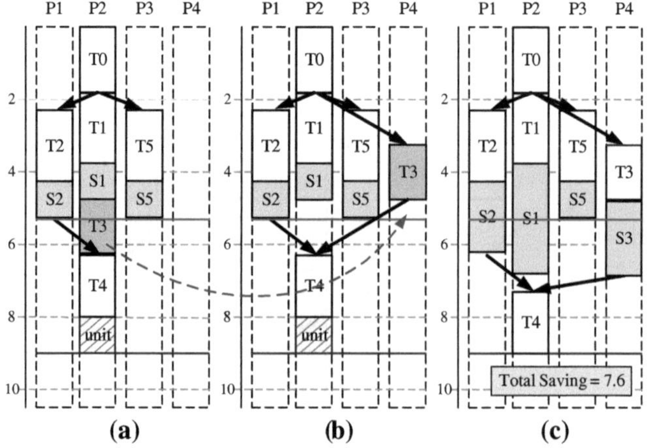

Fig. 5.4 Example of re-scheduling

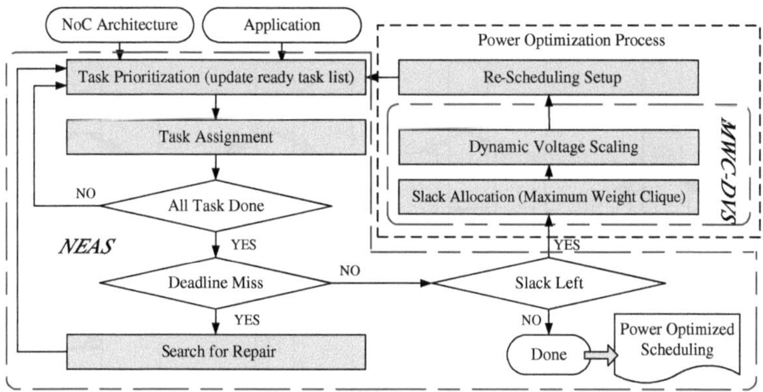

Fig. 5.5 Flow of our proposed energy-aware task scheduling algorithm for dynamic voltage scalable systems

predefined cost function. While all the tasks are scheduled, repair process proposed by Hu and Marculescu [2] will be triggered if there is any deadline constraint violation. Otherwise, if there is any left slack, our power optimization algorithm will be applied to find a set of tasks for slack allocation and to scale the operating voltage for energy saving. After the DVS process, we re-schedule the successive tasks below the *Re-Schedule_Hyper-Plan* to obtain a better result iteratively.

5.3.1 Task Prioritization

In the Task Prioritization process, we first find the budget slack for each task which depends on the energy consumption of the task on different PEs using the same

Task Prioritization

Input: current schedule
Output: the task with maximum priority
1. update Ready task list (RL)
2. for each task τ_k in RL, calculate $min_{f(\tau_k)} = min\{F(\tau_k, p_i) \mid \forall p_i \in P\}$
3. if $\exists\ min_{f(\tau_k)} - BD(\tau_k) > 0$, return τ_k which has maximum $min_{f(\tau_k)} - BD(\tau_k)$
4. for each task τ_k in RL, find minimum energy consumption E_1^k without violation
5. for each task τ_i in RL,
   schedule task τ_i to a PE that corresponds to minimum E_1^i
6. for each task τ_k in RL, k !=i, find $E_1^{'k}$ without violation
7. calculate the energy penalty function $\delta_E^i = \sum_{\forall k \in RL, k \neq i} (E_1^{'k} - E_1^k)$
8. remove the task τ_i from the schedule
9. return the task which has minimum value of energy penalty δ

Fig. 5.6 Procedure to prioritize task assignment

budget deadline (BD) calculation as in the EAS algorithm [2]. However, after BD allocation, we modify the cost function used in EAS for a more accurate prediction of energy consumption.

In the EAS algorithm, the metric $\delta_E^i = E_2^i - E_1^i$ is calculated for all tasks and the task in the ready task list (RL) which has the largest δ_E for task assignment will be selected. E_1^i and E_2^i are respectively the minimum and the second minimum energy (Due to length restrictions, the detailed description of EAS algorithm is not shown here). However, the above calculation of δ_E only compares the penalty of energy dissipation of τ_i itself in different PEs, we should further consider the impacts from tasks other than τ_i. Therefore, we modify the energy dissipation penalty metric as:

$$\delta_E^i = \sum_{\forall k \in RL, k \neq i} \left(E_1^{\prime k} - E_1^k\right),$$

where $E_1^{\prime k}$ represents the minimum energy dissipation of binding task τ_k to an appropriate PE, which will have minimum energy dissipation at next iteration, in case of selecting task τ_i for task assignment at current iteration. The detailed description of our Task Prioritization is illustrated in Fig. 5.6. Please note that the tasks in this process are only scheduled for calculating the energy penalty function δ. Thus, the schedule tables of both links and the PEs will be restored every time after δ is calculated.

5.3.2 Task Assignment

The objective of Task Assignment is to determine a PE for the task τ_i, selected at the Task Prioritization step, to assign. In order to deliberate more suitable PEs for

τ_i, the energy-aware cost function for assigning τ_i to $M(\tau_i)$ must include: computation energy $\left(E_{M(\tau_i)}^{v_{max}}\right)$ under maximum voltage v_{max}, communication energy $(\sum_{\forall j, vol(e_{i,j})\neq 0} E_{RA(p_i,p_j)})$, and potential energy saving $\left(E_{virtual\_DVS}^{M(\tau_i)}\right)$ that may be gained by the power optimization process. Such function is defined as follows:

$$E_{cost}^{M(\tau_i)} = E_{M(\tau_i)}^{v_{max}} + \sum_{\forall j, vol(e_{i,j})\neq 0} E_{RA(p_i,p_j)} + E_{virtual\_DVS}^{M(\tau_i)}$$

$$E_{virtual\_DVS}^{M(\tau_i)} = -S_{v_k,v_l}^{M(\tau_i)} \times \Delta t, \qquad \text{if } F(\tau_i, M(\tau_i)) < BD(\tau_i)$$

$$= \infty, \qquad \text{if } F(\tau_i, M(\tau_i)) \geq BD(\tau_i)$$

where $F(\tau_i, M(\tau_i))$ denotes the earliest finish time of τ_i at $M(\tau_i), S$ and Δt is the time difference between $\max\{F(\tau_i, p_i \in P)\}$ (that is also less than BD) and $F(\tau_i, M(\tau_i))$. The idea of potential energy saving is to increase the priority of the PE that has earlier $F(\tau_i, p_i)$ for assignment candidate, because the time difference Δt may be used for voltage scaling.

Figure 5.7 shows an example on the task assignment of τ_i at the intermediate stage. Assigning task τ_i to different PEs consumes different values of energy (communication and computation energy) as illustrated in Fig. 5.7a. In this case, we will discard P4 for assignment since assigning τ_i to P4 will cause BD violation $(F(\tau_i, M(\tau_i)) \geq BD(\tau_i))$. Figure 5.7b shows an example on the calculation of $E_{cost}^{M(\tau_i)}$, where we consider the $E_{virtual\_DVS}^{M(\tau_i)}$ for potential energy saving. The detailed procedure is shown in Fig. 5.8. P3 will be selected for assignment since it has the lowest value of $E_{cost}^{M(\tau_i)}$.

5.3.3 Power Optimization

To identify a suitable scaling voltage for a task with respect to its assigned PE to achieve the most efficient energy dissipation is a challenging DVS problem. Previous work on voltage scaling problem essentially examined the PE power variations based on the executed task and allocated the available slacks to one of the executed tasks at one time for energy saving as illustrated in the motivational example. In other words, if we can allocate one more unit time of slack to an executed task that has the most computational energy saving, such task will be selected for voltage scaling, and so on. However, locally considering energy saving at one task cannot promise the efficiency of slack distribution. In order to achieve better energy saving via voltage scaling, a better slack allocation methodology should consider energy saving in terms of per unit time of slack:

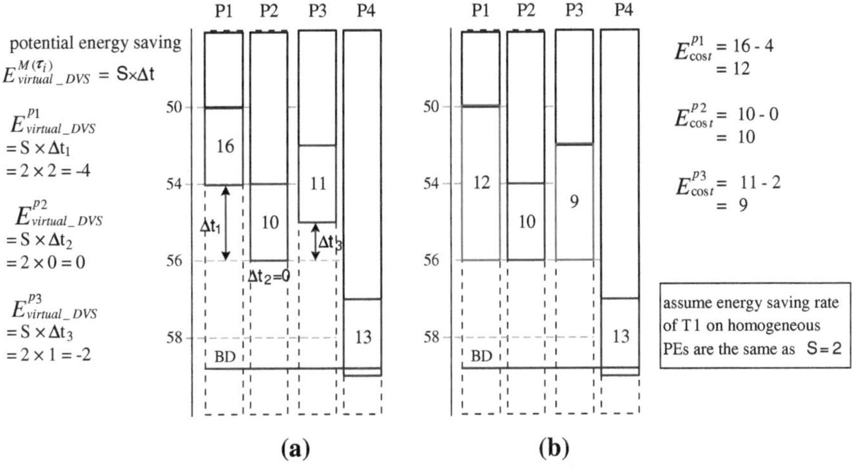

Fig. 5.7 Example of task assignment process

Task Assignment

Input: A task τ_i in RL selected by Task Prioritization process

Output: An assignment for τ_i with minimum cost of Energy ($E^{M(\tau_i)}_{cost}$)

1. for each PE (p_k), compute earliest finish time of τ_i as $F(\tau_i, p_k)$
   and check the BD violation of assigning τ_i to p_k
2. if all PE violate BD, return the PE which has minimum BD violation time
3. else, finish_time_max = $\max\{F(\tau_i, p_k) \mid \forall p_k \in P, F(\tau_i, p_k) < BD(\tau_i)\}$
4. for each p_i which is not violated
5. $\Delta T = finish\_time\_max - F(\tau_i, p_i)$
6. $E^{p_i}_{virtual\_DVS} = -S^{p_i}_{v_k,vl} * \Delta T$
7. $E^{p_i}_{cost} = E^{V_{max}}_{p_i} + \sum_{\forall j, vol(e_{i,j}) \neq 0} E_{RA(p_i, p_j)} + E^{p_i}_{virtual\_DVS}$
8. return the PE which has minimum value of $E^{M(\tau_i)}_{cost}$

Fig. 5.8 Procedure to assign a task to an appropriate PE

$$E_{per\_slack\_sav} = \frac{\sum_{\tau_i \in slack\_insert\_task} E_{saving}}{\sum expense\_slacks},$$

where the numerator denotes the total energy saving after slack allocation achieved by voltage scaling, and the denominator is the total expense of slacks by deadline node tasks.

Then, we reconsider the example in Fig. 5.3d. Extending one unit-time at T0 costs 2 unit-times of slack at deadline nodes T4 and T5 trading off with 1.2 units of energy saving. In other words, the per-slack energy saving $E_{per\_slack\_sav}$ in this case is 0.6 unit of energy. However, in the case of Fig. 5.4g, each slack can provide much more $E_{per\_slack\_sav}$ such as 1.4. Therefore, in this section, we will formulate the slack allocation problem into an MWC problem and introduce our

Table 5.1 Reference data for power optimization

Task	T0	T1	T2	T3	T4	T5	T6
S	1.1	1.2	0.7	0.7	1.0	0.9	0.6
Execution time	1.5	1.5	1.5	1.5	1.5	1.5	1.5
BD	2.5	5.0	7.5	7.5	6.7	6.7	10.0
Finish time	1.5	4.5	6.5	6.0	3.5	3.0	8.5
Extention_bound	1.0	0.5	1.0	1.5	3.2	3.7	1.5
Extendable_time	0.5	0.5	1.0	1.5	1.5	1.5	1.5

new voltage scaling algorithm which can efficiently optimize the power consumption during the scheduling process. Table 5.1 lists the reference data for the extension time calculation of all tasks used in this chapter.

Figure 5.9 shows an example of our power optimization process. At the initial stage, assuming the task graph in Fig. 5.9a has been mapped to P1, P2 and P3 and scheduled as illustrated in Fig. 5.9b. We can find that there are still 1.5 unit-times of slack left for power optimization. In order to obtain better energy saving, we want to allocate this slack to multiple tasks at the same time. To make sure that the distributed slack do not affect each other, the selected task should be located to different data dependent paths and executed at different PEs. In other words, for the selected set of tasks, extending the execution time at one task will not affect the ability of extension of other selected tasks. In addition to the original data dependent path as shown in Fig. 5.9a, dotted lines as shown in Fig. 5.9c represent the locality dependency in each PE since extending the execution time at a former task also affects the other tasks scheduled at the same PE afterwards. The two additional dependent paths are added and shown with dotted lines. All the tasks at the same dependent path can be seen as fully connected in the dependency graph as illustrated in Fig. 5.10a since each dependent path can only has one task to be selected for voltage scaling. This can be formed as an MWC problem to find the independent set of tasks and to extend its execution time.

In order to find the maximum energy saving, the complementary graph of Fig. 5.10a can be used as the input for the MWC problem as illustrated in Fig. 5.10b, where all the cliques that can be found and their weights are listed. The MWC of the graph $G_c = (V_c, E_c)$ is a sub-graph $G_{MWC} \in G_c$, where all vertices in G_{MWC} are pair-wise adjacent and the total weight of all the vertices in G_{MWC} is maximum. This is known to be an NP-complete problem [3, 4] and should be solved using a heuristic polynomial-time algorithm. The Cliquer tool [5], which is based on a branch-and-bound technique, is used in this chapter. We can find that the MWC is formed by T2, T3 and T4 in this example and we can insert slacks to these tasks for voltage scaling.

After task selection for voltage scaling, the next step is to decide the available extension time for this set of tasks. We used the reference data as listed in Table 5.1 for the extension time calculation of all tasks. Figure 5.11 shows the pseudo code on the *extendable_time* decision for each task. After the extendable time decision, a power optimization process is executed to find the *extension_time* for the set of tasks selected by the MWC tool, and then Re-Scheduling Setup procedure as described in Fig. 5.12 will be executed. In this case, the available

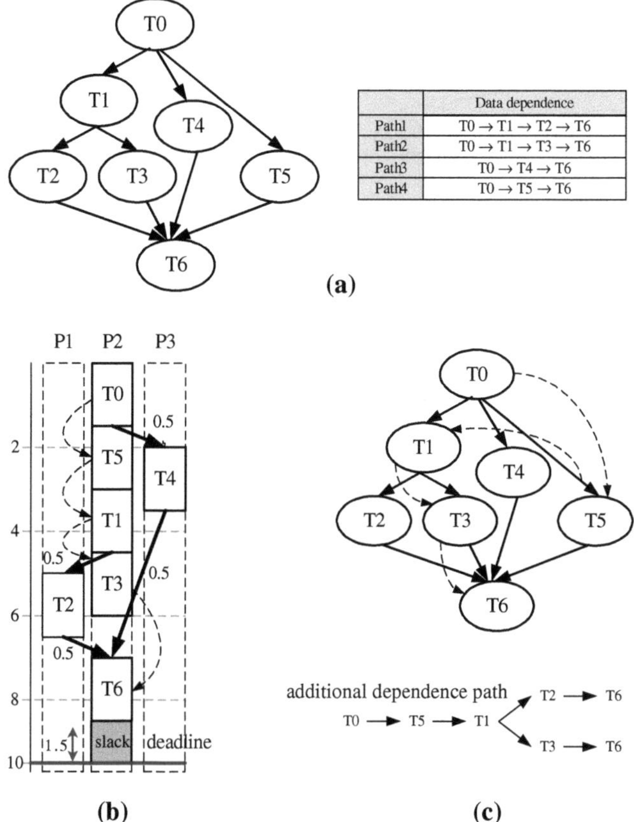

Fig. 5.9 Example for power optimization

extension times for T2, T3, and T4 are 1.0, 1.5, and 1.5 respectively. Therefore, in the next iteration of power optimization, T2, T3, and T4 are all extended 1.0 for voltage scaling. Since our algorithm promises the independence of the selected tasks for power optimization, in observation of Fig. 5.10c, we can find that the schedule in front of the selected tasks will not be affected by the extension of T2, T3 and T4. Therefore we can set a *Re-Schedule_Hyper-Plane* at the front of the selected tasks and divide the scheduled application into *Preserve Region* and *Re-Schedule Region*. Then, only the *Re-Schedule Region* should be rescheduled in power optimization process iteratively.

5.3.4 Re-Scheduling Setup

Since the voltage scaling process will extend the execution times for the tasks selected by the MWC tool, all the successive tasks will be delayed. However, if we

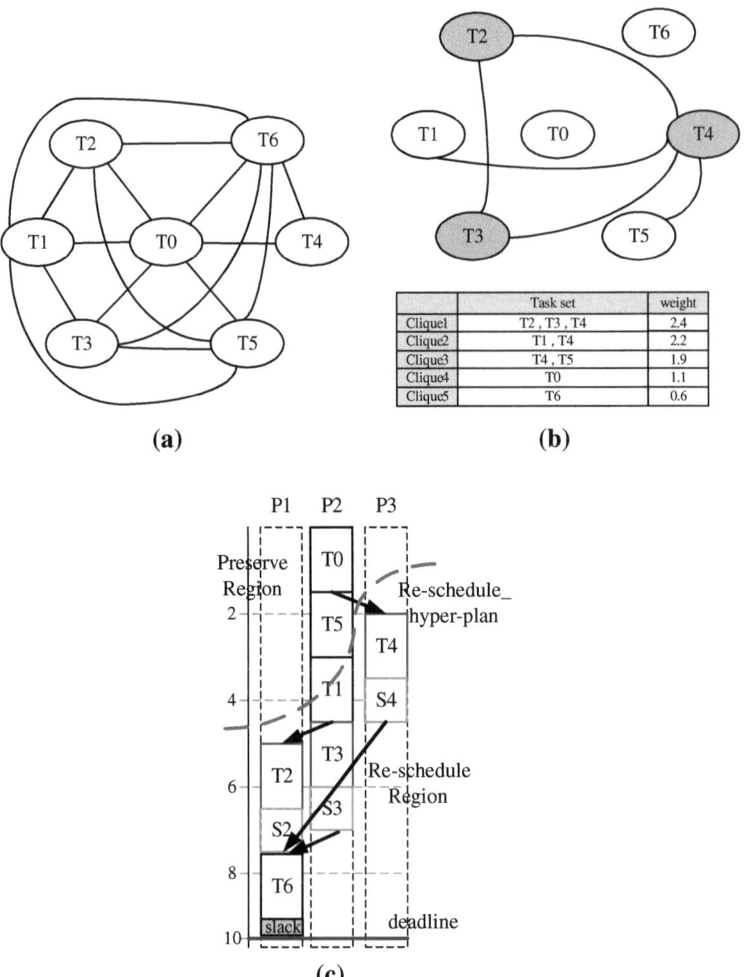

Fig. 5.10 MWC formulation example for power optimization

re-consider the motivational example in Fig. 5.4, we can find that re-scheduling task T3 to P4 essentially trades the communication overhead with the potential energy saving $E_{virtual\_DVS}^{M(\tau_i)}$, as illustrated in the cost function of the Task Assignment process. In this case, assigning T3 to P4 instead of P2 can get 1.62 (0.9×1.8) units of potential energy saving even with 1.0 unit of communication energy overhead. Therefore, integrating a re-scheduling process to our framework has the chance to flatten the executed tasks to multiple PEs and may increase the slack for potential energy saving. Figure 5.12 shows the Re-Scheduling Setup procedure which decides the Re-Schedule Region for next iteration of task scheduling.

```
=====================================================================
Power_Optimization
---------------------------------------------------------------------
Input: A generated schedule
Output: Preserve_Region
1. construct Directed Dependence Graph with locality and data dependence (DDG)
2. initialize an independent graph Gc=(Vc,Ec) where Vc=T, Ec= φ
3. for each pair of task τi and task τj , independent_build(τi, τj , DDG,Gc)
4. for each task τi, if ( extendable_time(τi)× S^{M(τi)}_{vk,vl} < Emin )
                          remove τi and connected edges from Gc
5. Maximum weight independent set (MWIS) = CLIQUER( independent graph Gc )
6. for all τi ∈ MWIS   // Re-Schedule_Hyper-Plan
7.         extention_time(τi) =  min{extendable_time(τk) | ∀τk ∈ MWIS}
8. for all τi ∈ MWIS , update the execution time and execution energy of τi
9. Re-Scheduleing_Setup(MWIS,DDG)
10. return schedule for new iteration // Preserve_Region
=====================================================================
independent_build( τi, τj, DDG,Gc)
---------------------------------------------------------------------
1. if there is an edge between vertex vi an vj in Gc, return true
2. if  τj ∈ susscesors(τi) in DDG, return false
3. else, for each τk ∈ susscesors(τi) ,
4.                    if (independent_build( τk,  τj) = false), return false
5.        construct an edge between vertex vi an vj in Gc
6.        return true
=====================================================================
extendable_time(τi)
---------------------------------------------------------------------
1. extension_bound = BD(τi) - finish_time(τi)
2. extenable_time = min{ extension_bound , extendable_time(τk) | ∀ τk ∈ successor (τi)}
3. return extendable_time
=====================================================================
```

Fig. 5.11 Procedure of our power optimization algorithm

```
=====================================================================
Re-Scheduling_Setup
---------------------------------------------------------------------
Input: Maximum Weight Independent Set (MWIS),
       Directed Dependence Graph with locality and data dependence (DDG)
Output: The schedule for next iteration    //return Preserve_Region
1. For each task τi ∈ MWIS  , Delete_Reschedule_Region( τi,DDG)
=====================================================================
Delete_Reschedule_Region( τi ,DDG)
---------------------------------------------------------------------
1. For each task τj ∈ susscesors(τi)  , Delete_Reschedule_Region(τj ,DDG)
2. remove τi from schedule
=====================================================================
```

Fig. 5.12 Procedure of the re-scheduling process

Table 5.2 Algorithms used for experiments

Algorithm	EAS	NEAS+PV-DVS	NEAS+PV-DVS+RES	NEAS+MWC-DVS	NEAS+MWC-DVS+RES
Scheduling	EAS	NEAS	NEAS	NEAS	NEAS
Voltage scaling	N/A	PV-DVS	PV-DVS	MWC-DVS	MWC-DVS
Re-Scheduling	N/A	X		X	

5.4 Experimental Results

To evaluate the effectiveness of our framework in energy saving, we conducted experiments on various task graphs generated by the standard package TGFF [6]. To obtain accurate on-chip communication estimation, a scalable mesh-based NoC architecture was used for performance analysis. We also applied the dimension-ordered XY routing restriction for communication traffic. In addition to our proposed NEAS+MWC-DVS+RES algorithm, we also implemented EAS algorithm and PV-DVS technique for comparison. Table 5.2 lists the combinations of algorithms used in this experiment. To clarify the benefit of solving DVS using MWC-based concept, NEAS+PV-DVS combining our proposed NEAS scheduling algorithm with PV-DVS technique was used for comparison. Both NEAS+PV-DVS+RES and NEAS+MWC-DVS+RES were added with a re-scheduling process to further improve energy saving.

As illustrated in Table 5.3, ten task graphs (each has about 100 tasks with a vast amount of dependencies as listed in column 2) were used for simulation on an NoC fabric that consisted of a 6 × 6 mesh network of routers. For dynamic voltage scaling, we assumed that all PEs are homogeneous and can be operated at two voltage levels such as 1.2 and 0.8 v under 0.13 μm technology with v_t value of 0.4 v. For fair comparison, all the energy values were normalized to the result of EAS algorithm and the energy saving improvement percentage toward NEAS+PV-DVS were listed individually. First of all, we can find that NEAS+PV-DVS had about 11.68% of energy saving with respect to EAS algorithm since EAS did not apply any voltage scaling technique. However, comparing NEAS+PV-DVS with our proposed NEAS+MWC-DVS, we can find that our proposed technique achieved even 9.03% more energy saving improvement with respect to NEAS+PV-DVS since the MWC-based slack allocation approach can find the set of tasks with the most energy saving in each iteration. Furthermore, NEAS+PV-DVS+RES demonstrated the benefit of re-scheduling since the energy saving improvement was increased to 63.35% with respect to NEAS+PV-DVS. Moreover, after joint consideration with re-scheduling, our proposed NEAS+MWC-DVS+RES algorithm achieved even more energy saving improvement to 97.34%.

To examine the slack utilization performance with respect to energy saving. Figure 5.13a shows the energy saving when we extend the deadline restriction for the same TGFF pattern as listed in Table 5.3. The y-axis represents the percentage of the extended deadline to the original deadline. The energy consumption results

Table 5.3 Energy saving comparison with various algorithms

TGFF category	Node/ edge	EAS	NEAS+PV-DVS		NEAS+PV-DVS+RES		NEAS+MWC-DVS		NEAS+MWC-DVS+RES	
		Energy (%)	Energy (%)	Saving (%)	Energy (%)	Saving improve (%)	Energy (%)	Saving improve (%)	Energy (%)	Saving improve (%)
TGC0	103/259	100	89.09	10.91	75.74	109.79	88.55	3.94	71.10	193.17
TGC1	112/140	100	94.70	5.30	86.95	63.75	94.33	10.44	84.15	185.59
TGC2	111/326	100	94.83	5.16	90.46	8.34	94.35	−0.89	88.44	19.52
TGC3	103/173	100	93.86	6.14	88.40	67.80	91.85	33.21	86.03	90.15
TGC4	109/162	100	82.67	17.33	66.07	93.41	81.00	10.15	63.68	108.03
TGC5	101/143	100	91.19	8.81	77.90	171.74	91.39	−0.91	77.68	174.78
TGC6	109/208	100	79.00	21.00	77.74	−2.36	76.08	13.98	72.48	27.71
TGC7	103/227	100	89.30	10.70	84.78	29.63	88.79	4.16	83.01	47.89
TGC8	106/185	100	80.56	19.44	79.89	5.51	79.25	6.89	76.30	23.75
TGC9	102/190	100	87.87	12.13	78.72	85.89	86.72	9.37	76.68	102.80
Average		100	88.32	11.68	80.67	63.35	87.23	9.03	77.96	97.34

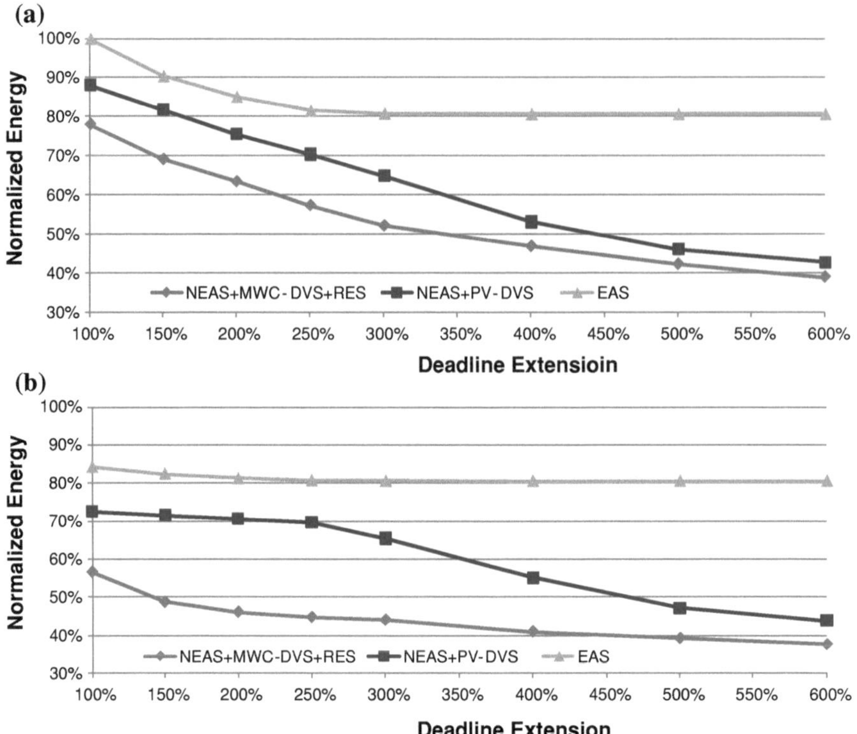

Fig. 5.13 Energy consumption of different algorithms under various deadline constraints

of ten TGFF patterns were averaged and normalized to the EAS algorithm running at the original deadline restriction generated by TGFF tool for comparison. We can find that the energy consumption of EAS algorithm did improve when extension time was large since it did not apply any DVS technique. On the other hand, our proposed NEAS+MWC-DVS+RES achieved more energy saving than NEAS+PV-DVS since our proposed slack allocation methodology did efficiently use the slack for voltage scaling especially when there were more slacks left for use. As the deadline extension percentage increases, the performance of the two algorithms will become closer since most of the tasks will be scheduled to a small number of PEs and running at the lowest voltage level. Therefore, the main energy consumption difference of the two algorithms was the communication overhead that can be minimized by our energy cost function, $E_{cost}^{M(\tau_i)}$, which traded communication overhead with potential energy saving during the re-scheduling process.

To demonstrate the benefit of re-scheduling, we reduced the total communication traffic to one-fifth of the original TGFF benchmark and illustrated the energy results in Fig. 5.13b. The energy consumption offset represents the lowered communication overhead. We can find that the energy consumption difference

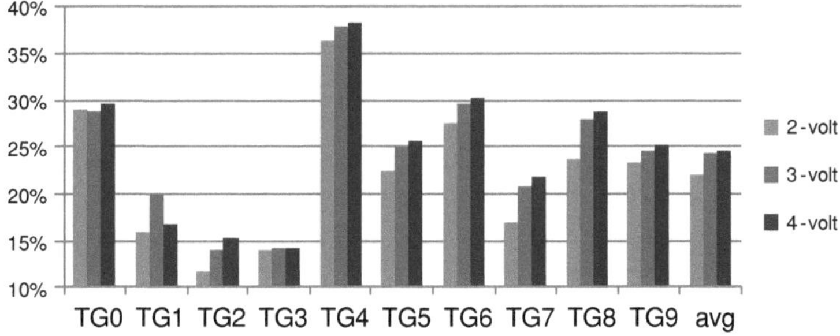

Fig. 5.14 Energy saving percentage versus different available voltage levels

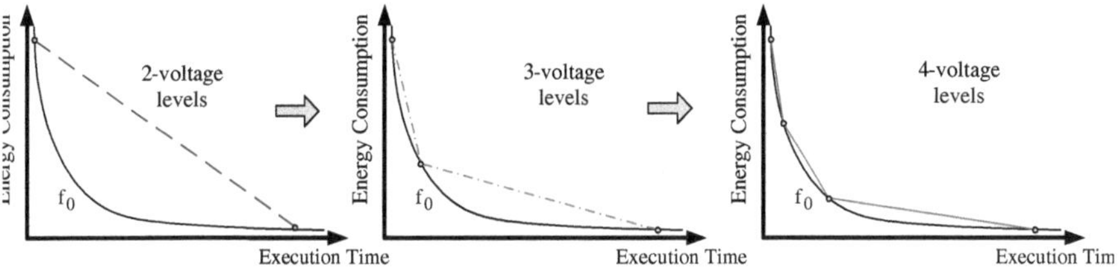

Fig. 5.15 Energy consumption versus execution time under different voltage levels

between NEAS+PV-DVS and NEAS+MWC-DVS+RES was larger than the result in Fig. 5.13a, since the opportunity for trading communication overhead with potential energy saving was increased. In other words, our algorithm provided more chance to flatten the executed tasks to multiple PEs and increase the slack for potential energy saving.

In the previous experiment, each PE only used two voltage levels for scaling. Figure 5.14 shows the energy saving percentage when we used 3 v (1.2, 1.0, and 0.8 v), and 4 v (1.2, 1.0, 0.9, and 0.8 v) for comparison. We can find that the energy saving increased since the real minimum energy consumption that can be reached approached the ideal curve f_0 as illustrated in Fig. 5.15 as the number of available voltage levels increased.

5.5 Remarks

In this chapter, we proposed an energy-aware task scheduling algorithm for NoC-based dynamic voltage scalable system which can efficiently allocate the unused slacks for energy saving by voltage scaling. For energy-aware scheduling, our proposed NEAS algorithm considered the potential energy saving at the Task Prioritization and Task Assignment stages in advance. For power optimization, our proposed MWC-DVS algorithm formulated the slack allocation problem as a maximum weight clique problem to find a set of tasks for scaling voltage in a greedy way and for saving energy iteratively. For better slack utilization, an efficient re-scheduling technique to adjust the scheduling along with a power optimization process was proposed. Experimental results using TGFF standard benchmarks showed that our algorithm efficiently utilized the slack for energy saving compared with the conventional algorithms.

References

1. T. Ishihara and H. Yasuura, "Voltage Scheduling Problem for Dynamically Variable Voltage Processors", in *Proceedings of the International Symposium on Low Power Electronics and Design*, pp. 197–202, August 1998
2. J. Hu and R. Marculescu, "Energy-Aware Communication and Task Scheduling for Network-on-Chip Architetures under Real-Time Constraints", in *Proceedings of the Conference on Design, Automation and Test in Europe*, pp. 234–239, February 2004
3. M. R. Garey and D. S. Johnson, *Computers and Intractability: a Guide to the Theory of NP-Completeness*, Freeman and Company, 1979
4. I. M. Bomze, M. Budinich, P. M. Pardalos, and M. Pelillo, *The Maximum Clique Problem,* Kluwer Academic, 1999
5. S. Niskanen, *Cliquer*, http://users.tkk.fi/pat/cliquer.html, Accessed July 2011
6. R. Dick, D. Rhodes, and W. Wolf, "TGFF: Task Graphs for Free", in *Proceedings of the International Workshop on Hardware/Software Codesign*, pp. 97–101, March 1998

Part III
Case Study: Bidirectional NoC (BiNoC) Architecture

Chapter 6
Bidirectional Noc Architecture

A Bidirectional channel Network-on-Chip (BiNoC) architecture is proposed in this chapter to enhance the performance of on-chip communication. In a BiNoC, each communication channel allows itself to be dynamically reconfigured to transmit flits in either direction. This added flexibility promises better bandwidth utilization, lower packet delivery latency, and higher packet consumption rate. Novel on-chip router architecture is developed to support dynamic self-reconfiguration of the bidirectional traffic flow. The flow direction at each channel is controlled by a channel-direction control protocol. Implemented with a pair of finite state machines, this channel-direction control protocol is shown to be high performance, free of deadlock, and free of starvation.

6.1 Problem Description

In a conventional NoC architecture, each pair of neighboring routers uses two unidirectional channels in opposite direction to propagate data on the network as shown in Fig. 6.1a. In our BiNoC architecture, to enable the most bandwidth utilization, data channels between each pair of routers should be able to transmit data in any direction at each run cycle. That is, four kinds of channel-direction combinations should be allowed for data transmission as shown in Fig. 6.1b. However, current unidirectional NoC architectures, when facing applications that have different traffic patterns, cannot achieve the high bandwidth utilization objective.

Note that the number of bidirectional channels between each pair of neighboring router in BiNoC architecture is not limited to two. The more bidirectional channels that can be used, the better the performance results. However, in order to provide a fair comparison between our BiNoC and the conventional NoC that usually provided two fixed unidirectional channels for inter-router communication, only two bidirectional channels were used in BiNoC as illustrated in Fig. 6.1.

S.-J. Chen et al., *Reconfigurable Networks-on-Chip*,
DOI: 10.1007/978-1-4419-9341-0_6,

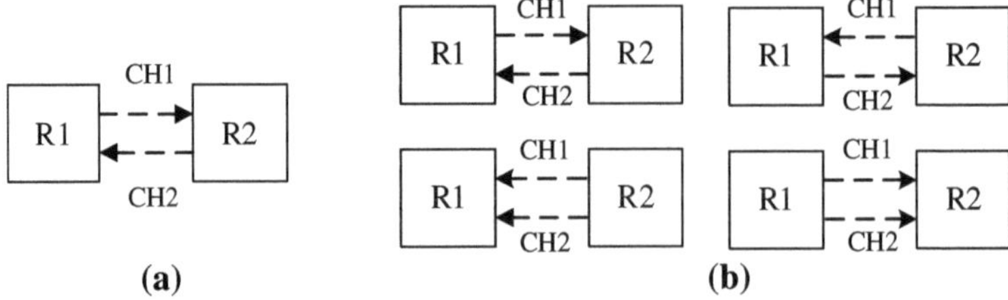

Fig. 6.1 Channel directions in a typical NoC and proposed BiNoC

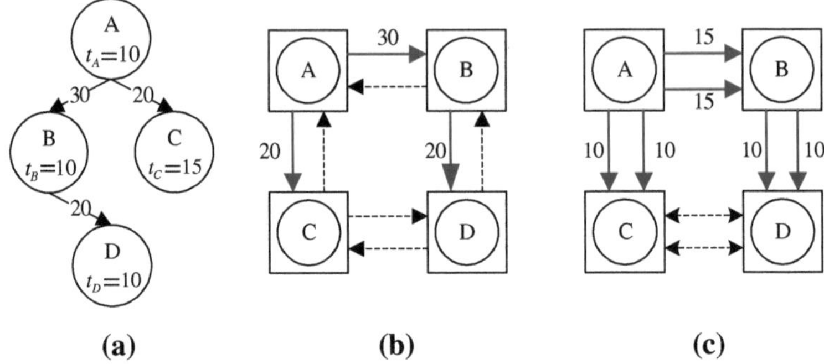

Fig. 6.2 Example of task graph mapping on typical NoC and BiNoC

6.1.1 Motivational Example

As shown in Fig. 6.2a, an application task graph is typically described as a set of concurrent tasks that have already been assigned and scheduled onto a list of selected PEs. Each vertex represents a task with a value t_j of its computational execution time and each edge represents the communication dependence with a value of communication volume which is divided by the bandwidth of a data channel.

For the most optimized mapping in a 2×2 2-D mesh NoC as shown in Fig. 6.2b, the conventional NoC architecture in this case can only use three channels during the entire simulation and result in a total execution time of 80 cycles. However, if we can dynamically change the direction of each channel between each pair of routers like the architecture illustrated in Fig. 6.2c, the bandwidth utilization will be improved and the total execution time be reduced to 55 cycles. Figure 6.3 shows the detailed execution schedules, where the required communication time between nodes in BiNoC is extensively reduced.

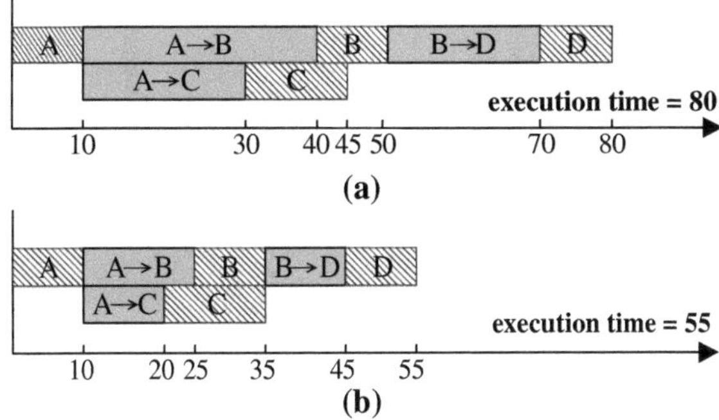

Fig. 6.3 Detailed execution schedules of typical NoC and BiNoC

6.1.2 Channel Bandwidth Utilization

During the execution of an application, the percentage of time that a data channel is kept busy is defined as channel bandwidth utilization U. To be more specific,

$$U = \frac{\sum_{t=1}^{T} N_{\text{Busy}}(t)}{T \times N_{\text{Total}}},$$

where T is the total execution time, N_{Total} is the total number of channels available to transmit data, and $N_{\text{Busy}}(t)$ is the number of channels that are busy during clock cycle t. It is obvious that $U \leq 1$.

We have developed a cycle-accurate NoC simulator to evaluate the performance of a given NoC architecture. Additional implementation details of this NoC simulator will be elaborated in later sections. Using this simulator, we measured the channel bandwidth utilizations of a conventional NoC with respect to three types of synthetic traffic patterns: uniform, regional, and transpose. The channel utilization against different traffic volumes are plotted in Fig. 6.4 under both XY and Odd–Even routing.

Figure 6.4a, b plot the bandwidth utilizations of a conventional NoC router with virtual-channel flow-control. Four virtual-channel buffers, each with a depth of 8-flits, are allocated in each flow direction. Figure 6.4c, d give the percentage of time that exactly one channel is busy and another channel is idle among time intervals when there is at least one channel busy. Figure 6.4e, f give the percentage of time that a bidirectional channel may help alleviating the traffic jam when exactly one channel is busy and the other is idle. Figure 6.4a, c, e results are obtained using XY routing; and Fig. 6.4b, d, f use Odd–Even routing.

From Fig. 6.4a, b, it is clear that even with the most favorable uniform traffic pattern, the channel bandwidth utilization peaks under XY routing and Odd–Even routing are only around 45 and 40% respectively under heavy traffic. For the transpose traffic pattern under XY routing, which is considered the worst case scenario, U falls even below 20%. In other words, in a unidirectional channel setting, even with two channels between a pair of routers, at most one channel is

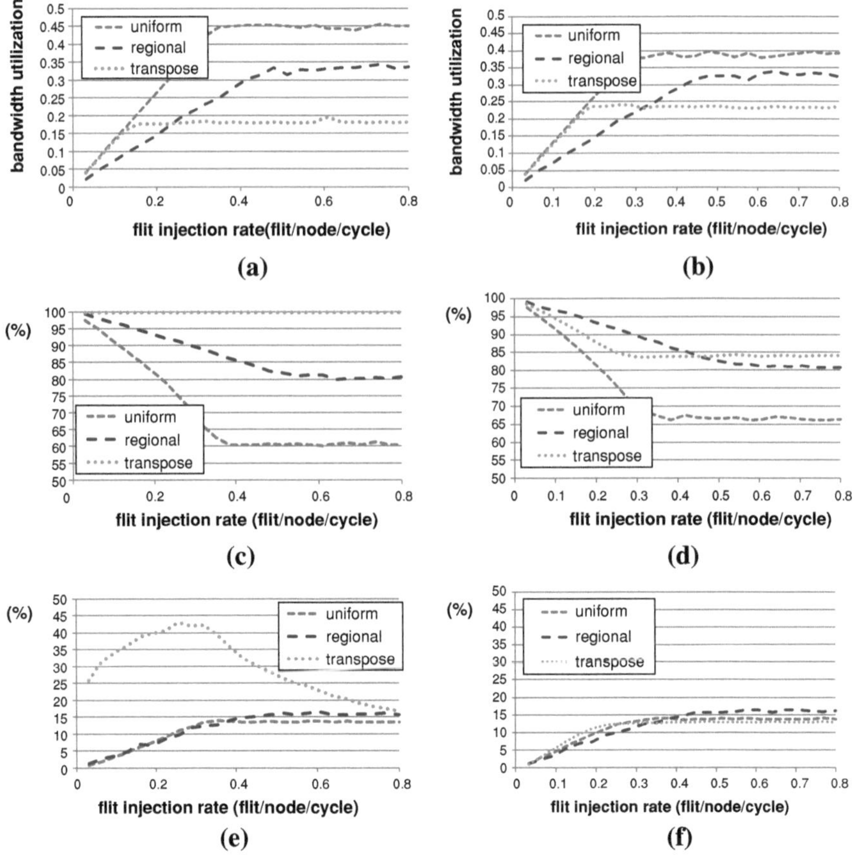

Fig. 6.4 Bandwidth utilizations analysis of a conventional NoC router

kept busy on average during normal NoC operation despite the deterministic routing algorithm such as XY or adaptive routing algorithm such as Odd–Even.

One possible cause of the low bandwidth utilization as shown in Fig. 6.4a, b is due to few bottleneck channels that take too long to transmit data packets in the designated direction. To validate this claim, we examine how often both channels between a pair of router are kept busy simultaneously. In Fig. 6.4c, d, the percentage of time that exactly one channel is busy and the other is idle given that one or both channels are busy is plotted respectively under XY and Odd–Even routings. As traffic load increases, it is clear that a significant amount of traffic utilizes only a single channel while the other channel is idle.

However, the situation where one channel is busy and the other is idle could be the case where there are no data that need to transmit in the opposite direction of the busy channel. It does not reveal whether there are additional data packets waiting in the same direction as the busy channel. These data packets are potential candidate to take advantage of the idle channel if the idle channel's direction can be reversed. In Fig. 6.4e, f, the percentage of time, that there are data packets needed to transmit along the same direction as the busy channel, while the other channel remains idle out of all situations where exactly a single channel is busy is plotted. An important observation is that for large traffic volume, this situation happens for about 15% of time *despite the type of traffic patterns or the routing methods*.

Figure 6.4 gives ample evidence that the unidirectional channel structure of current NoC cannot fully utilize available resources (channel bandwidth), and may cause longer latency. This observation motivates us to explore the BiNoC architecture that offers the opportunity to reverse channel direction dynamically to relieve the high traffic volume of a busy channel in the opposite direction.

6.2 Bidirectional Channel

NoC is a comprehensive on-chip data communication infrastructure that involves many design issues. Approaches such as packet routing technique, application mapping, scheduling, and topology synthesis, have been explored for the purpose of improving NoC performance [1–7].

However, considering the physical level of a backbone, the interconnect wire between routers is also an important factor in determining the total performance of a system. To the best of our knowledge, this is the first work to analyze and promote the use of bidirectional wires for NoC interconnection.

6.2.1 Design Requirements

Bidirectional channels have been incorporated into off-chip multi-processor high-speed interconnecting subsystems over years. Recently, bidirectional on-chip global inter-connecting subsystems have also been studied quite extensively for supporting electronic design automation of system-on-chip platforms [8–11]. Hence, physical layer design of an NoC channel to support bidirectional data transmission should be of little difficulty. The real challenge of embracing a bidirectional channel in an NoC is to devise a distributed channel-direction control protocol that would achieve several important performance criteria:

1. *Correctness*: It should not cause permanent blockage of data transfer (deadlock, starvation) during operation.
2. *High Performance*: Its performance should be *scalable* to the size of the *NoC* fabric and *robust* with respect to increasing traffic volume. In addition, it is desirable that the performance enhancement can be achieved across different characteristics of application traffic patterns.

3. *Low Hardware Cost*: The hardware overhead to support the bidirectional channel should be small enough to justify the cost-effectiveness of the proposed architecture.

6.2.2 Related Works

Faruque et al. [12] presented a configurable on-chip communication link named 2X-Links to support bidirectional transmission using tri-state logic. The major difference between their work and the BiNoC reported in this book is how the channel-direction decision is made. Essentially, the direction decision mechanism proposed in [12] is centralized and each link should be configured in advance according to the off-line analysis result based on application bandwidth requirements. Beside, the configuration process is handled by FPGA. On the other hand, the channel-direction decision schemes of our proposed BiNoC architecture is distributed in each router and can be dynamically self-reconfigured according to real-time traffic need; and every channel-direction decisions are made at individual routers in a distributed manner. All reconfigurations are decided by the inter-router channel-direction control protocol and handled within the BiNoC fabric without any outside intervention (*e.g., FPGA board*).

Another bidirectional channels NoC architecture is reported [13]. The main difference between [13] and this work is the channel-direction control protocol. In [13], a pressure-based channel-direction control protocol is proposed. This is markedly different from the acquisition-based channel-direction control protocol proposed in this book. A side-by-side comparison between these two protocols is included in this book. It is shown that the acquisition-based channel-direction control protocol is provably free of the type of deadlock suffered by the pressure based protocol reported in [13].

6.3 BiNoC: Bidirectional NoC Router Architecture

In the following sections, we will introduce two flow-control mechanisms including wormhole and virtual-channel flow-control for BiNoC architectures. The implementation details will be described and breakdown into function blocks for easier understanding.

6.3.1 BiNoC Router with Wormhole Flow-Control

To realize a dynamically self-reconfigurable bidirectional channel NoC architecture, we initially modified the input/output port configuration and router control

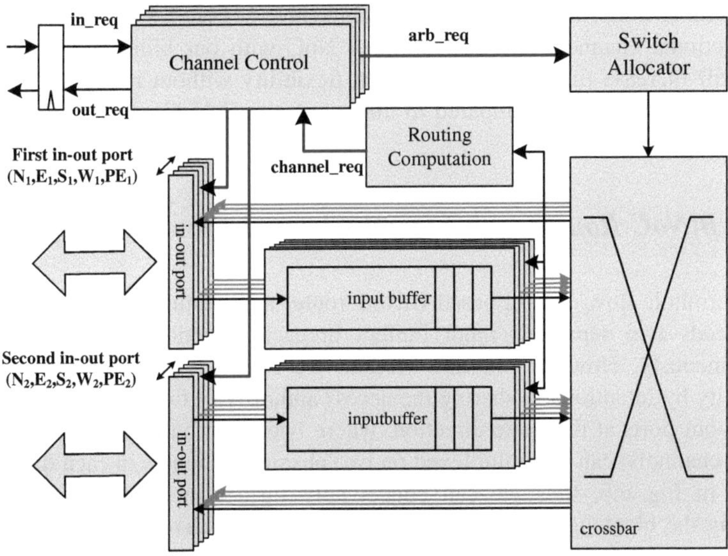

Fig. 6.5 Proposed BiNoC router with wormhole flow-control

unit designs based on the conventional router using wormhole flow-control as we proposed in [14]. In order to dynamically adjust the direction of each bidirectional channel at run time, we add a *Channel Control* module to arbitrate the authority of the channel direction as illustrated in Fig. 6.5.

Each bidirectional channel which is composed of an in–out port inside is the main difference from the conventional router design where unidirectional channel employs a hardwired input port or output port. However, the total number of data channels is not changed as its applicable bandwidth for each transmission direction is doubled.

In our design, each channel can be used as either an input or an output channel. As a result, the width of a channel request signal, *channel_req*, generated from the RC modules is doubled. Two bidirectional channels can be requested in each output direction. In other words, this router is able to transmit at most two packets to the same direction simultaneously which decreases the probability of contentions.

The channel control module has two major functions. One is to dynamically configure the channel direction between neighboring routers. Since the bidirectional channel is shared by a pair of neighboring routers, every transition of the output authority is achieved by a channel-direction control protocol between these two routers. The control protocol, implemented as FSMs, which will be described in the following section. The other responsibility is that whether the channel request (*channel_req*) for the corresponding channel is blocked or not will depend on the current status of channel direction. If the channel is able to be used, the *arb_req* will be sent to the SA to process the channel allocation.

The most important point of this architecture is that we can replace all the unidirectional channels in a conventional NoC with our bidirectional channels. That will increase the channel utilization flexibility without requiring additional transmission bandwidth compared to the conventional NoC.

6.3.2 BiNoC Router with Virtual-Channel Flow-Control

The wormhole flow-control based BiNoC router architecture that we proposed in [14] needs two separated input buffers in each direction to receive packets simultaneously. However, in this section, we improve the channel utilization flexibility by intentionally sharing the access authority of two input buffers for the two in–out ports at the same direction, where two input buffers (or even multiple virtual-channels) can be multiplexed on two physical channels in each direction as shown in Fig. 6.6, thus we can conceivably further increase performance by reducing the blocking effect of links. Since the virtual-channels in each direction can be shared by the two physical channels simultaneously in BiNoC, the total number of virtual-channels is equal to that in a conventional virtual-channel flow-control based router.

The concept of sharing virtual-channels is analogous to a conventional virtual-channel flow-control based router design but increases the channel utilization flexibility since the virtual-channels in each direction can be shared by the two physical channels simultaneously in BiNoC. Since the virtual-channel buffers are shared at the same direction, the total number of virtual-channel buffers is equal to that in a conventional virtual-channel flow-control based router design.

As shown in Fig. 6.6, our virtual-channel flow-control based BiNoC is implemented with a 4-stage pipeline architecture to enhance the clock rate and throughput as a conventional router design does. To fit in with our BiNoC architecture, the proposed channel control models as described in the following sections will be also implemented to arbitrate the inter-router bidirectional channel authority. Note that all the input/output ports in a router are registered to provide clean signals from router to router in our design, as an NoC hardware implementation may require long wires.

6.3.3 Reconfigurable Input/Output Ports

As shown in Fig. 6.6, one of the input/output ports is designated as a high-priority (HP) port, and the other is designated as a low-priority (LP) port. Each of the two bidirectional channels between a pair of routers will determine its own transmission direction based on a distributed channel-direction control protocol. When both channels have the same transmission direction, two data packets could be sent concurrently, which effectively double the channel bandwidth.

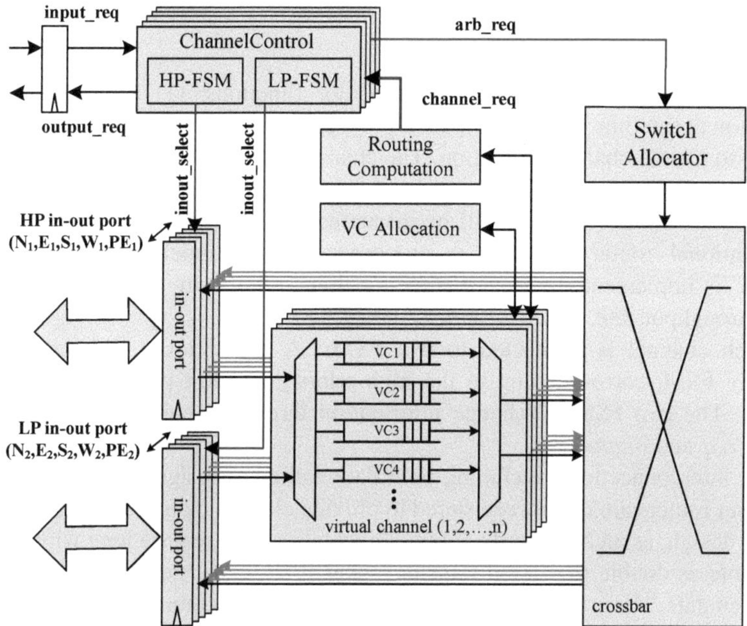

Fig. 6.6 Proposed BiNoC router with virtual-channel flow-control

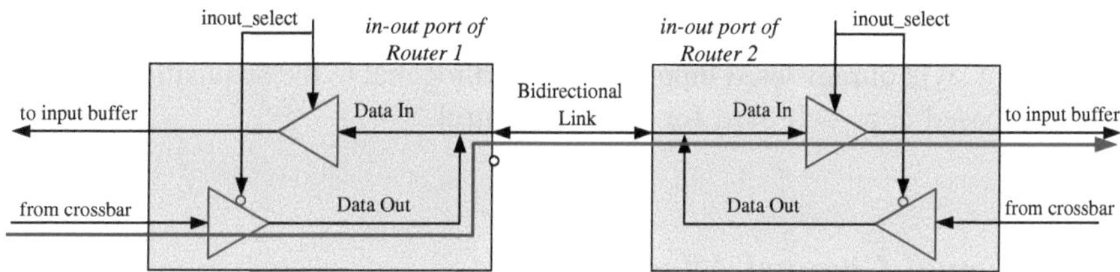

Fig. 6.7 Schematic of bidirectional link implemented in BiNoC

Figure 6.7 shows the detailed schematic of the in–out ports implementation in BiNoC. As long as the *inout_select* signals are assigned properly, no conflict and no unpredictable situation will occur. Instead of using dedicated input port and output port, within BiNoC, each port can be either an input port or an output port controlled by the *inout_select* signals generated from the *Channel Control* module.

Faruque et al. [12] and Cho et al. [13] also used tri-state logic to realize bidirectional transmission on an NoC platform. Our tri-state logic, as illustrated in Fig. 6.7 differs from these works in the control logic, controlled by a FSM implemented in the *Channel Control* module. This flexible design of tri-state logic enables the bidirectional transmission on a single wire and improves the bandwidth utilization without additional wires. The hardware cost of this kind of tri-state logic is quite reasonable and will be discussed in the following section.

6.3.4 Channel Control Module

The channel control module has two major functions: (1) determine channel direction at run time; and (2) output an *arb_req* signal to the switch allocator at the router to handle channel allocation. The channel control module is realized with a finite state machine (FSM) consisting of three states. Details of the operations of the channel control module will be discussed in the next section. Similar to the conventional router design, our proposed virtual-channel flow-control based BiNoC is implemented with a 4-stage pipeline architecture to enhance the clock and throughput rate.

Each channel is connected to two FSMs: A high-priority FSM and a low-priority FSM, corresponding to the high-priority and low-priority ports of each router. The two FSMs exchange information through a pair of signaling wires: *input_req* and *output_req*.

All interconnections, including both data and control signals between the two adjacent routers are doubly registered to provide clean signals from router to router in our design, as an NoC hardware implementation may require long wires. This is desirable as double registered transfer makes it easier to meet the timing closure requirements. Nevertheless, the round trip delay while communicating between a pair of routers is increased to 4 cycles accordingly. In other words, the depth of each buffer in our design should be at least 4-flits long to ensure no data overflow occurs. This design style, however, also incurs additional latency during the channel-direction control process. If these registers are removed as indicated in [13], then the channel-direction reversal process should only incur one dead cycle. The design style of registered input/output is orthogonal to the acquisition-based or pressure-based approach used for channel control.

6.3.5 Virtual-Channel Allocator

The virtual-channel allocator (VA) module matches resource requests from input virtual-channels to available virtual-channels at downstream routers. Peh and Dally has detailed the complexity of a general VA in [15]. Since the virtual-channel allocator matches resource requests from input virtual-channels with available output virtual-channels, its hardware overhead increases with the number of virtual-channels used in each direction. The complexity and latency of a VA depends on the range of the routing selection. According to our NoC design, the routing circuit returns the candidate output virtual-channels in a single physical channel to be selected by an input virtual-channel. The request size of the input arbitration is equal to the number of virtual-channels, v, sharing a physical channel. With the concerns of circuit area and critical path delay, we chose a basic separable allocator proposed in [15] to perform virtual-channel allocation as illustrated in Fig. 6.8.

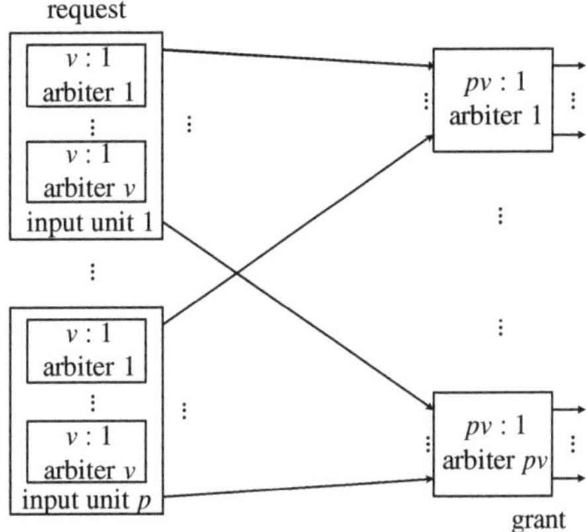

Fig. 6.8 A virtual-channel allocator in a BiNoC router

Given a routing result calculated by the RC module, VA needs the first-stage arbiter at each input virtual-channel to select one requested virtual-channel at the downstream router. Then, at the second-stage, the request contentions between input virtual-channels will be resolved as illustrated in Fig. 6.8. Totally, we need *pv* *v*:1 arbiters at the input side and *pv* *pv*:1 arbiters at the output side. This VA architecture is identical to the conventional virtual-channel flow-control based router [15, 16]. Since we share the virtual-channel buffers among the two bidirectional channels in each direction, there is no overhead in the VA stage compared to the conventional virtual-channel flow-control based router design.

6.3.6 Switch Allocator

The switch allocator (SA) allocates a time slot of the cross-bar switch to move flits from an input virtual-channel to an output physical channel. In a conventional NoC, SA is accomplished with a two-stage arbitration, where the arbiter has an equal number of input lines and output lines, and its output is the grant signal as discussed in [15].

In our BiNoC architecture, the number of channel bandwidth available in each output direction is doubled from *p* to 2*p*. The activity of each output arbiter is controlled by the corresponding channel state. The output arbiter is enabled to process the requests and generates a grant only if the associated channel state is capable to output data. Otherwise, the output arbiter will be turned off. On the input side, in order to keep up with the double bandwidth obtainable in each output direction, the number of the first-stage arbiters is also doubled to be capable of picking two requests in each input direction, as illustrated in Fig. 6.9.

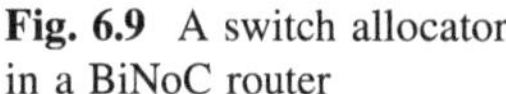
Fig. 6.9 A switch allocator in a BiNoC router

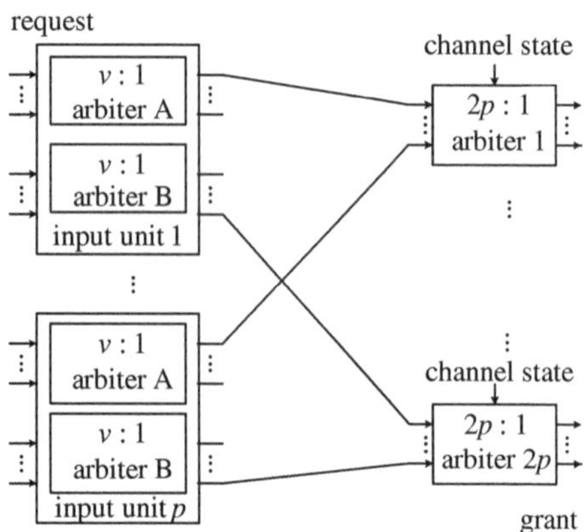

6.4 Bidirectional Channel Direction Control

To achieve the objective of self-reconfiguring the direction of a bidirectional channel dynamically, each bidirectional channel needs a pair of control schemes communicating to each other to arbitrate the channel authority. This chapter will describe in detail the bidirectional channel control scheme and its design mechanism. Starvation avoidance arbitration and deadlock free routing techniques will be introduced in the following sections.

6.4.1 Inter-Router Transmission Scheme

The data transmission direction of a bidirectional channel needs to be self-configured, dynamically determined at real-time based on local traffic demands. To achieve this goal, we propose an inter-router data transmission scheme as shown in Fig. 6.10, which illustrates the details of control blocks and signals for the two bidirectional channels between a pair of adjacent routers.

Configuration of a bidirectional channel direction is controlled by a pair of FSMs in the channel control blocks of the routers at both ends. To enhance routing efficiency and prevent starvation, one of the two FSMs will be designated with a higher priority (HP) and the other with a lower priority (LP), which will be described in detail in the following sections. To enforce fairness of data transmission, the FSMs on the other channel between the same pair of routers will be designated with opposite priorities. The operations of adjacent FSMs are fully synchronized against a common clock.

The two FSMs exchange control signals through a pair of doubly-buffered hand-shaking signals: *input_req* (input request) and *output_req* (output request). We have *output_req* = 1 when the sending-end router has data packet to transmit. The *output_req* signal from one router becomes the *input_req* signal to the FSM of

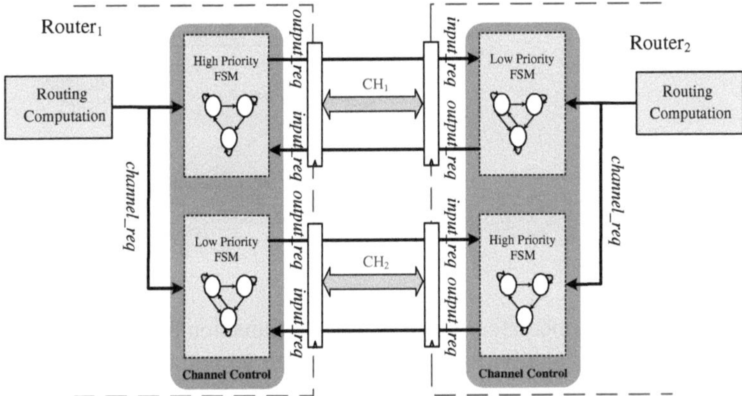

Fig. 6.10 Inter-router data transmission scheme

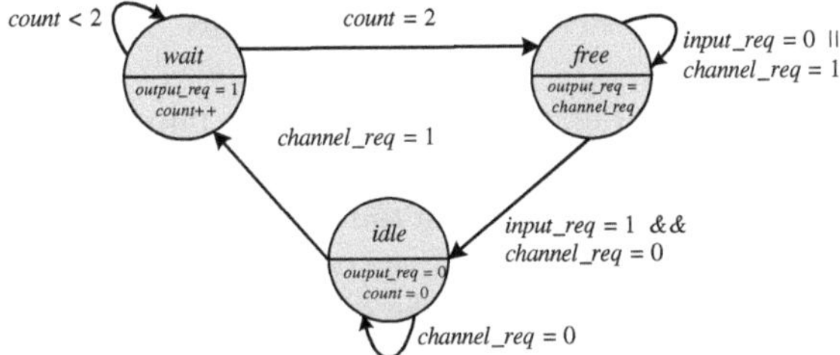

Fig. 6.11 FSM for HP port of bidirectional channels

the other router after two clock cycles due to the presence of two buffering flip-flops. Each FSM also receives a *channel_req* (channel request) signal from the internal RC module. We have *channel_req* = 1 when a data packet in the local router is requesting the current channel for forwarding data. However, if the downstream input buffer is full, *channel_req* will be reset to 0. Each FSM also has an internal counter to keep track of the propagation delay through the synchronous channel.

6.4.2 Bidirectional Channel Routing Direction Control

The state transition diagrams of a high-priority and a low-priority channel-direction control FSMs are respectively shown in Figs. 6.11 and 6.12. The interactions between these two adjacent routers via their corresponding FSMs are demonstrated in Fig. 6.10. Note that each FSM consists of three states: *free, wait,* and *idle* which are defined as follows:

1. *free state*: The channel is available to *output* data to the adjacent router.
2. *idle state*: The channel is ready to *input* data from the adjacent router.

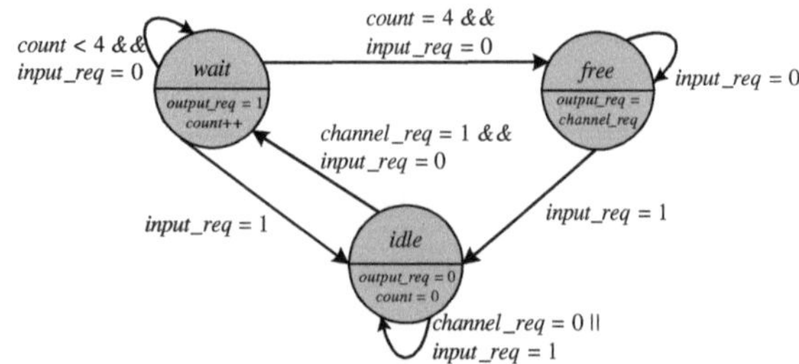

Fig. 6.12 FSM for LP port of bidirectional channels

3. *wait state*: An intermediate state preparing the transition from the *idle* state with an input channel direction to the *free* state with an output channel direction.

The detailed operations of the HP FSM and the LP FSM are discussed below.

6.4.2.1 High-Priority FSM Operations

As shown in Fig. 6.11, the HP FSM is initiated at the *free* state. It will remain in this state if *input_req* = 0 or *channel_req* =1. In other words, as long as there are data packets within current router to be sent via this channel, the channel direction will remain out-bound. Even when there is no data to transmit, if there is no request to send data from the other router, the channel direction should still remain unchanged. The only condition that the HP FSM will leave the *free* state and enter an *idle* state is when there is no data to transmit from the current router, AND there is data to be sent from the adjacent router. While at the *free* state, the output signal *output_req* = *channel_req*. Hence the output value may become 0, allowing the other router to request the channel.

Once the FSM enters the *idle* state, it will remain at that state as long as there is no data to be transmitted outbound (*channel_req* = 0). Moreover, while in an *idle* state, *output_req* = 0. As soon as *channel_req* = 1, the HP FSM will enter a *wait* state waiting to regain the channel control to transmit data.

At the *wait* state, *output_req* = 1 and a cleared counter will increase during each clock cycle in the *wait* state. As soon as *count* = 2, the HP FSM returns to the *free* state and starts data transmission. Meanwhile, the counter is reset to *count* = 0. The purpose of the *wait* state is to allow the *output_req* = 1 signal to reach the adjacent router such that the LP FSM will yield the channel by entering its *idle* state.

6.4.2.2 Low-Priority FSM Operations

As shown in Fig. 6.12, the LP FSM is initiated at the *idle* state with *output_req* = 0. It will leave the *idle* state for a *wait* state if the HP FSM of the other router yields the channel (*input_req* = 0) AND a local RC module requests to use the channel (*channel_req* = 1).

Being a lower priority-end of data transmission, the LP FSM will remain in the *wait* state for 4 clock cycles. During any of these four cycles, if the HP FSM requests the channel (*input_req* = 1), the LP FSM will return to an *idle* state. Only after four cycles (*count* = 4) and if *input_req* = 0, the LP FSM will enter a *free* state, starting transmission.

However, different from the HP FSM, an LP FSM may remain in the *free* state only if HP FSM does not have any data to transmit (*input_req* = 0). Once *input_req* = 1, the LP FSM will cease data transmission immediately and fall back to an *idle* state. Therefore, as soon as *input_req* = 1, the LP FSM will enter the *idle* state regardless of its current state.

6.4.2.3 Channel Authority Confliction

With two FSMs, there is a total of 9 possible channel state combinations. However, since the channel can only assume one transmission direction at each clock cycle, the state combination of (*free, free*) must not occur. Here the first state refers to the state of the HP FSM and the second state is that of the LP FSM. Our proposed channel-direction control protocol can promise that there is no channel authority confliction between neighboring routers as follows:

If HP FSM is initiated to the free state and the LP FSM to the idle state, then the state transition diagrams specified in Figs. 6.11 and 6.12 *guarantee that the forbidden state combination* (*free, free*) *will never occur during normal operations.*

To elaborate that the pair of HP and LP FSMs cannot enter into a joint state of (*free, free*), one only needs to examine three other joint states (*wait, free*), (*free, wait*), and (*wait, wait*). According to Figs. 6.11 and 6.12, there are no other joint states that may make a state transition into the (*free, free*) joint state.

Case I Current joint state is (*wait, free*). According to Fig. 6.11, the output of HP FSM is *output_req* = 1. If the *channel_req* signals for both FSMs remain unchanged, *input_req* will be logic 1 two cycles after *output_req* changes to logic 1 at the HP FSM. According to Fig. 6.12, the LP FSM will change its state from *free* to *idle*. Thus, even the HP FSM will move from *wait* to *free,* the LP FSM will move from *free* to *idle*. As such, it is not possible to move from (*wait, free*) to (*free, free*).

Case II Current joint state is (*free*, *wait*). Hence the output of LP FSM is *output_req* = 1. This implies that after two cycles, the input of HP FSM will be *input_req* = 1. For the LP FSM to remain in the *wait* state, it is required that the LP FSM's *input_req* = 0. This implies that the HP FSM has *channel_req* = 0 for consecutive cycles. Thus, the HP FSM will enter into an *idle* state prior to the LP FSM entering a *free* state. Hence, the legitimate next joint state will be (*idle*, *free*).

Case III Current joint state is (*wait, wait*). In this case, the outputs *output_req* of both HP and LP FSMs will be set to logic 1. After 2 cycles, they turn into *input_req* = 1. For the HP FSM, it will ignore this input and enter the *free* state; while for the LP FSM, it will retreat to the *idle* state.

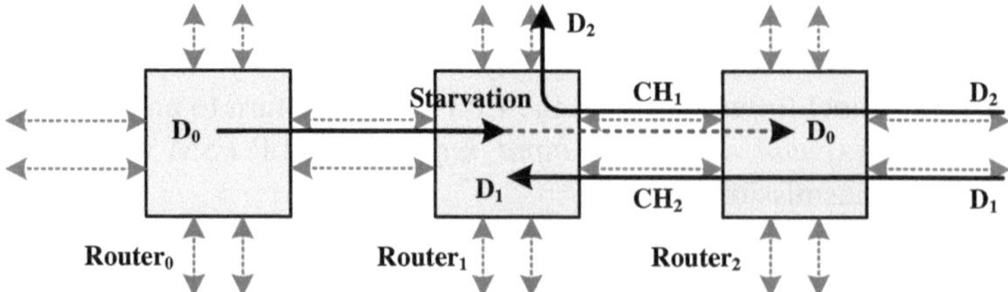

Fig. 6.13 Example of possible condition of starvation in BiNoC

Since none of these prior joint states can make a transition into the (*free, free*) state, the theorem is proved.

6.4.3 Resource Contention

While traffic is traversing on the network, there are two important resource contentions that may cause either a starvation or a deadlock problem. In the following sections, we will describe our arbitration technique in detail which can efficiently avoid the aforementioned problems.

6.4.3.1 Inter-Router Starvation

The bidirectional channel-direction control protocol presented in the previous section is a prioritized approach with one end designated a higher priority and the other a lower priority. Hence even the demands for a channel from both ends are about the same, the HP end will be given more cycles to transmit data, while the LP end will get less. One concern is that the LP end may be *starved* as a result of deprivation of its *fair share* of channel usage time. To illustrate, consider Fig. 6.13 where $Router_2$ is the HP end and $Router_1$ is the LP end for both channels CH_1 and CH_2. As such, D_1 and D_2 may occupy both channels over an extended period of time, causing starvation to the east-bound data package D_0.

To address this concern of channel starvation, we propose to assign opposite priorities to each of the pair of channels. For example, as illustrated in Fig. 6.14, one may assign $Router_1$ as the HP end for channel CH_1; while assign $Router_2$ as the HP end for channel CH_2. As a result, the east bound data packet D_0 will be assigned with channel CH_1 without further delay, over-writing the demand of D_2 as illustrated in Fig. 6.15. Meanwhile, D_1 and D_2 will compete for the same channel CH_2 for data transmission in the same manner as in a conventional uni-directional NoC architecture.

Intuitively, with one HP end at each side, the available data transmission capacity at each channel will be the same. Hence, it is unlikely that one side can

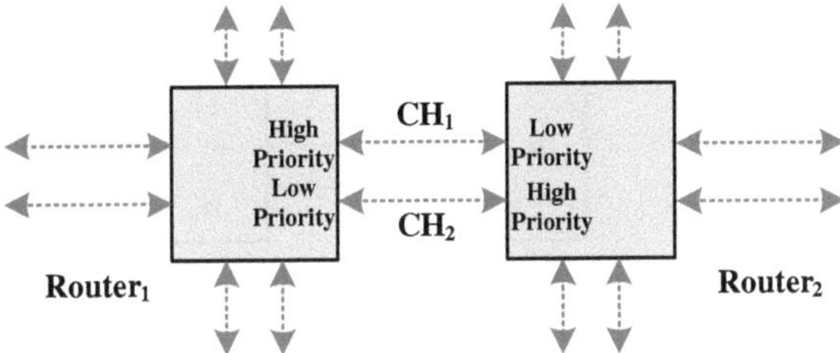

Fig. 6.14 Fixed channel prioritization mechanism

dominate the data transfer direction. Moreover, this proposed channel-direction control protocol can be interpreted as a generalization of the conventional unidirectional NoC. In particular, if one always sets the state of the HP end FSM as *free*, and the LP end FSM as *idle*, then the starvation problem can be eliminated.

Note that the side benefit of prioritized channel direction also reduces the channel-direction switching probability that may cause additional dead cycles. Since the prioritized approach regulates the traffic travel in opposite directions using different bidirectional channel, the inter-router channel contention is avoided thus dead cycles are reduced.

6.4.3.2 Deadlock-Free Routing

Deadlock occurs in an interconnection network when a group of *agents* are unable to make progress because they are waiting on one another to release *resources*. The four necessary conditions for deadlock to occur were first described by Owens et al. [17]. Essentially, if a sequence of waiting *agents* forms a cycle, the network is deadlocked. Consequently, deadlock can be avoided if the circular waiting relationship is broken in the resource dependence graph of an interconnection, because a cycle in the resource dependence graph is a necessary condition for deadlock.

In our BiNoC architecture, even though we have changed the physical channel characteristics, we still follow the flit-buffer flow-control design policies. Therefore, as clarified in [16], the *agents* and *resources* involved in deadlock condition are still packets and buffers (or virtual-channels) respectively under both wormhole flow-control and virtual-channel flow-control. In other words, the bidirectional physical channels between neighboring routers will not change the dependence relationship of *resources* in a BiNoC architecture and keeps the same wait-for and hold relations while deadlock occurs. Figure 6.16a, b show the physical channel difference between a conventional router and our proposed BiNoC with two virtual-channels in each direction.

Take the two buffers (or virtual-channels) at east side of the left router ($R1._{E(v1)}$ and $R1._{E(v2)}$) as an example, we can find that the resource dependences for these two virtual-channels in the conventional NoC and our BiNoC are both $\{R2_{E(v1,v2)}, R2_{N(v1,v2)}, R2_{S(v1,v2)}\} \rightarrow R1_{E(v1)}$ and $\{R2_{E(v1,v2)}, R2_{N(v1,v2)}, R2_{S(v1,v2)}\} \rightarrow R1_{E(v2)}$

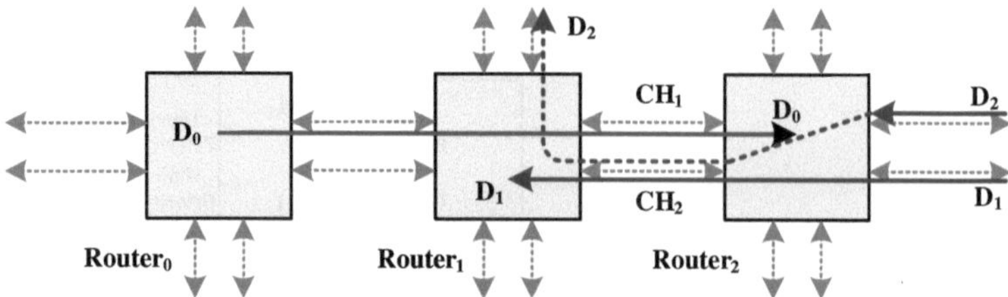

Fig. 6.15 Demonstration of starvation avoidance

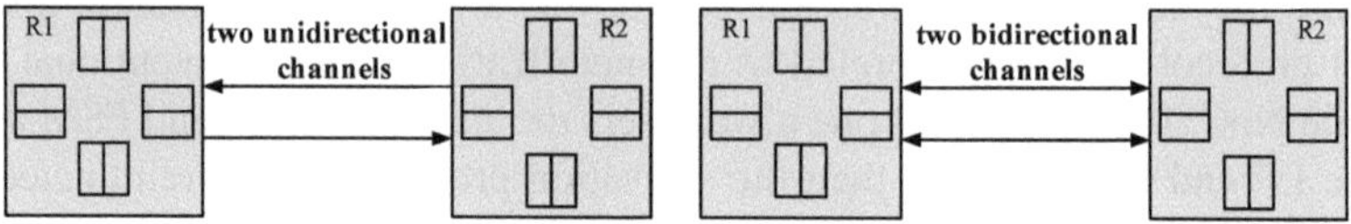

Fig. 6.16 Physical channel directions in **a** typical NoC and **b** BiNoC

as illustrated in Fig. 6.17. Similar dependency can also be found for buffers of $R1_{N(v1,v2)}$, $R1_{W(v1,v2)}$ and $R1_{S(v1,v2)}$ with buffers located at neighboring routers. Note that the resource dependence (→) is a transitive relation. If a → b and b → c, then a → c. Thus, we can find that the resource dependence graphs of Fig. 6.16a, b are the same.

For a deadlock to occur, *agents* are required to acquire some *resources* and wait on others in a manner that they will generate a cycle in the wait-for graph [18]. Thus, many existing deadlock avoidance routing techniques, such as dimension-ordered and the turn model based routing algorithms used in a conventional NoC architecture, can also be applied to our BiNoC.

In [13], it has been reported that under the pressure-based channel-direction control protocol, deadlock may occur. To illustrate the important distinction between the proposed acquisition-based channel-direction control protocol in BiNoC and the pressure based approach, the example reported in [13] will be discussed.

Figure 6.18 considers a situation that a flow f_B travels from node B to node C via node A, and all links connecting A with B are configured in the direction B → A. Assume another lighter flow f_A starts at D and heads toward B. Using pressure based channel-direction control protocol as proposed in [13], flow f_A may not exert enough pressure on the A → B link to overcome the pressure of f_B. As such, f_A may be blocked (illustrated by the open switch between node B and node A) even f_B cannot move forward. This blockage is likely to back fill the previous route segments. In particular, in the direction from node D to node C, both flows f_A and f_B will share the two channels in the same direction. It is possible that the blocked flow f_A may fill the buffer along its path and block other flows including f_B. Consequently, flow f_B will be blocked between node D and node C. This blockage sooner or later may in turn block both channels from node B and

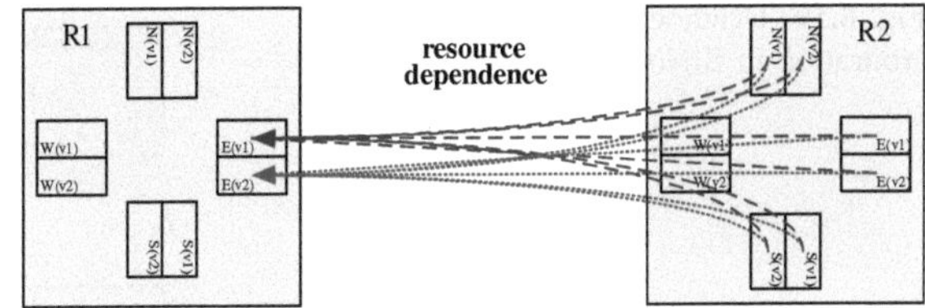

Fig. 6.17 Resource dependence relationship of east VC

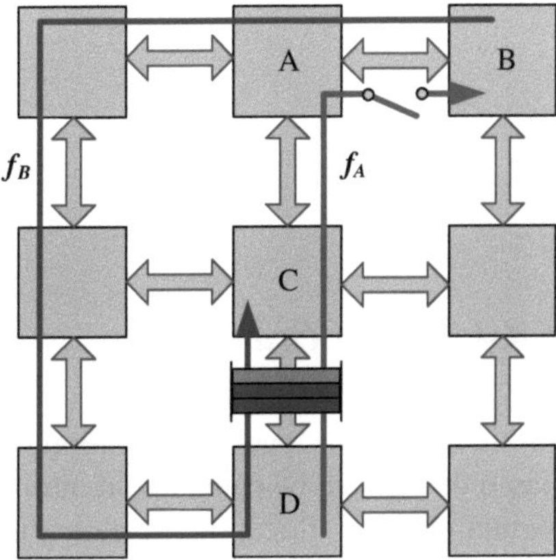

Fig. 6.18 Possible deadlock condition due to pressure-based design

node A. As a result, flow f_A is waiting flow f_B to complete and yield channels from node B to node A. But flow f_B will be waiting for flow f_A to yield buffer from node D to node C. In other words, a deadlock occurs.

Note that the deadlock arises only because the bidirectional nature of the link between A and B can cause the connection $A \rightarrow B$ to disappear due to pressure-based mechanism. In [13], a potential solution to this kind of deadlock problem is suggested which requires to ration available bandwidth among different flows according to their corresponding pressures. In other words, the two channels between nodes A and B will be configured with opposite directions in this case. If there are four bidirectional channels and f_B has higher pressure than f_A, pressure-based architecture will assign three channels to f_B and one to the f_A. However, the bottleneck of the circular dependence of f_A and f_B, at the A → B link, can only provide one channel to relieve the traffic in this case.

In contrast, in BiNoC, flow f_A will be able to gain both bidirectional channel accesses between node A and node B since the heavier flow f_B in the opposite direction is blocked due to limited buffer space at node A; thus the channel request from f_B will be disabled as illustrated in Fig. 6.19. In other words, it is useless to allocate physical channel to f_B. Thus, f_A can be delivered by all physical channels between nodes A and B, which can help to relief congestion faster. Until there is any available buffer space for f_B to move forward, the channel direction will be reallocated for both f_A and f_B based on the real-time traffic condition.

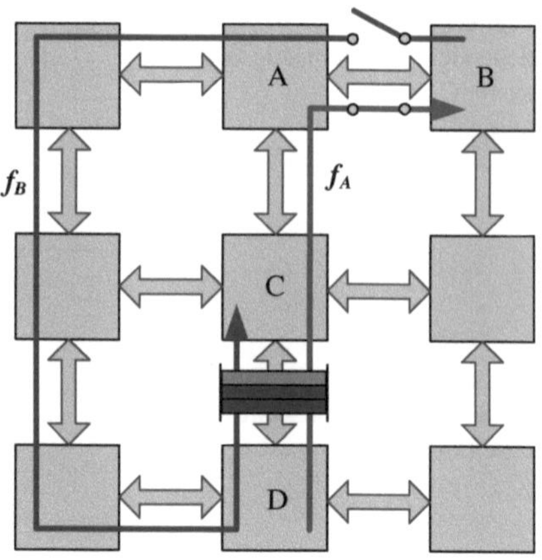

Fig. 6.19 Deadlock-free arbitration in BiNoC

6.4.4 Packet Ordering

In summary, the packet allocation methodologies for multiple virtual-channels based design can be categorized into static virtual-channel allocation and dynamic virtual-channel allocation. In static virtual-channel allocation, all packets from the same traffic flow use only a specific virtual-channel in each router, which ensures each flow can only use a single virtual-channel in each router thus guarantees packets being relieved in order. However, efficient static virtual-channel allocation requires a priori knowledge of the application's traffic pattern, which is unrealistic for general-purpose NoC design.

In dynamic virtual-channel allocation, available virtual-channels can be allocated to packets as they arrive, which provides more flexibility for virtual-channel utilization. However, two packets from the same flow may be assigned to different virtual-channels, which results in an out-of-order packet delivering. Many researches talking about guaranteed in-order packet delivery based on virtual-channel allocation technique or routing strategies such as [19–21] are outside the scope of our research. In this book, we guarantee in-order delivery of flits in the same packets but assume that each packet is an individual traffic flow with no ordering issue.

Even though the direction of two physical links in each direction can be dynamically changed, whether a packet will over-take a previous packet on the same connection (router) is resulted from the arbitration taken at the virtual-channel allocation stage. This problem will also happen in typical virtual-channel flow-control based NoC design with unidirectional physical links. Therefore, the research, on how to guarantee in-order packet delivery or packet reordering, can be seen as an orthogonal problem with dynamic channel-direction control protocol.

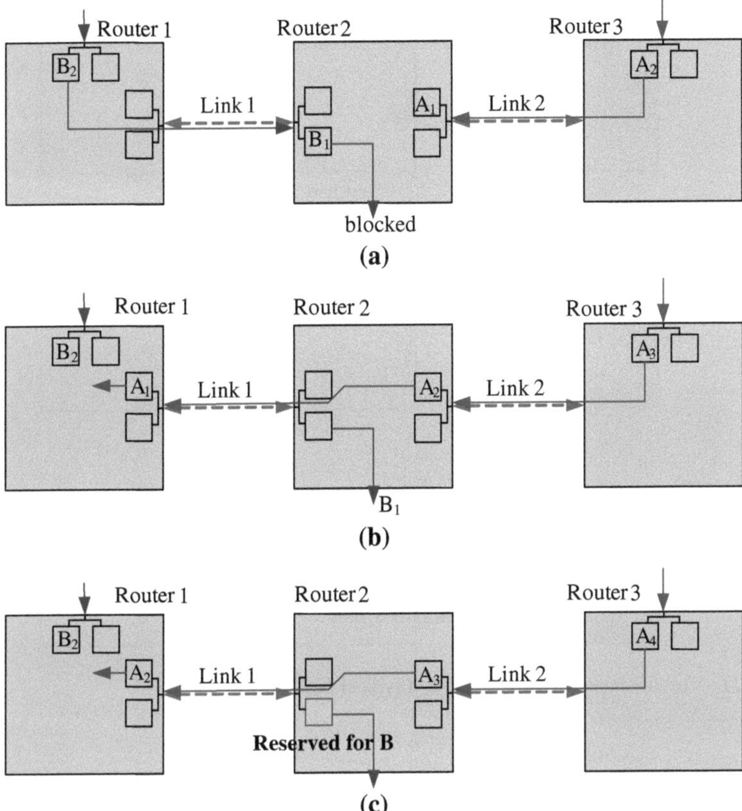

Fig. 6.20 Virtual-channel flow-control in BiNoC

6.4.5 Packet Transmission Interruption

The case that a packet is split over two links is a general case for virtual-channel flow-control based NoC architecture no matter its channel direction fixed or bidirectional. To clearly describe this problem, Fig. 6.20 illustrates the situation in our BiNoC architecture. To simplify the problem, we assume that there are two virtual-channels in each direction and one bidirectional link between each pair of routers.

We can find when flit B_1 is blocked at *Router_2* as shown in Fig. 6.20a, the authority of *Link_1* will be released. Thus, in this situation, the request of flit A_1 will reconfigure the *Link_1* direction for its transmission as shown in Fig. 6.20b. As a result, flit B_2 at *Router_1* may still be blocked while the congestion of flit B_1 at *Router_2* is already relieved and thus flit B_1 can be transferred without waiting for flit B_2 at *Router_1* as illustrated in Fig. 6.20c. Since flit B_1 is not the tail flit of packet B, the authority of virtual-channel in *Router_2* on its path will be reserved for flit B_2 but cannot be used by other packets.

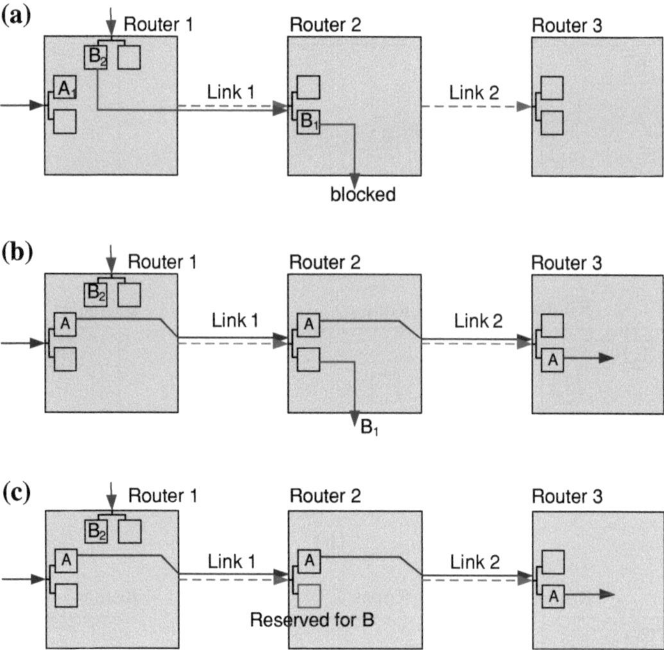

Fig. 6.21 Virtual-channel flow-control in a typical NoC

As for a typical unidirectional NoC, a packet may still be split under virtual-channel based flow-control as illustrated in Fig. 6.21. While flit B_1 is blocked at *Router_2* as illustrated in Fig. 6.21a, the authority of *Link_1* will be transferred to packet A. Then, flit B_1 may leave while flit B_2 is still blocked, and the virtual-channels on its path will be reserved for flit B_2.

Therefore, the case of packet split is not a problem resulted from link direction but from virtual-channel utilization methodology. If we lock the usage of *Link_1* for packet B in previous case, packet B will not be split but packet A will be blocked even if *Link_1* is idle. Therefore, to increase the link utilization, most typical virtual-channel flow-control methods only lock the virtual-channel authority to packet and leave the physical link authority for flexible usage.

6.5 BiNoC Characterization

Extensive simulation experiments using both synthetic and real world traffic patterns have been performed on BiNoC and their results are shown in this chapter. The results demonstrate clear performance advantage of BiNoC over that of traditional unidirectional NoC. In addition, implementation overhead in terms of area

Table 6.1 Area breakdown of different NoC architectures

Architecture	Buf. size/ dir. (flits)	Buf. depth (flits)	Ch./dir.	Crossbar	Freq. (MHz)	Normalized cycle
T-NoC_WH(32)	32	32	1-in 1-out	5 × 5	1,166	0.64
T-NoC_4VC(32)	32	8	1-in 1-out	5 × 5	746	1
T-NoC_4VC(64)	64	16	1-in 1-out	5 × 5	746	1
T-NoC_4VC(32)_4L	32	8	2-in 2-out	10 × 10	632	1.18
BiNoC_WH(32)	32	16	2-inout	10 × 10	921	0.81
BiNoC_2VC(32)	32	16	2-inout	10 × 10	666	1.12
BiNoC_3VC(48)	48	16	2-inout	10 × 10	637	1.17
BiNoC_4VC(32)	32	8	2-inout	10 × 10	627	1.19
BiNoC_4VC(64)	64	16	2-inout	10 × 10	561	1.33

and power will be described in detail in the following sections. Extensive experimental results shows that our proposed BiNoC architecture can provide better performance result with reasonable implementation costs.

6.5.1 Experiments Setup

The detailed simulation environments are described in Appendix A and the router architectures used for comparison are listed in Table 6.1.

In Table 6.1, T-NoC_4VC(32) represents a typical unidirectional NoC equipped with an input buffer size of 32 flits in each direction, which was divided into four 8-flit buffer queues used as virtual-channels. For each pair of neighboring routers, one input channel and one output channel were used to transmit data. Besides, T-NoC_WH(32) represents a typical unidirectional NoC architecture with wormhole flow-control which occupied one 32-flit buffer queues in each direction. Moreover, T-NoC_4VC(32)_4L, which was equipped with four unidirectional links between adjacent router pairs, was implemented to evaluate the effect of doubling inter-router communication bandwidth.

For our proposed BiNoC architectures, two bidirectional channels were used in each pair of neighboring routers. In order to inspect the performance trend among different buffer configurations and switching strategies, we implemented both wormhole and virtual-channel flow-control based router architectures. BiNoC_WH(32) denotes our proposed BiNoC architecture employing wormhole flow-control. According to our wormhole flow-control based BiNoC architecture described in [14], two input buffers were needed in each direction of BiNoC_WH(32), thus each inout port had one 16-flit input buffer. As for the virtual-channel flow-control, our proposed BiNoC architectures were configured with different numbers of virtual-channels such as BiNoC_2VC(32), BiNoC_3VC(48), and BiNoC_4VC(64) to respectively provide two, three, and four virtual-channels with 16 flits in depth in each direction. Note that the lengths of each virtual-channel in the above-mentioned three BiNoC architectures were all 16-flits in depth. To understand the effect caused by buffer depth, BiNoC_4VC(32) was also implemented with four virtual-channels in depth of 8 flits. Since buffer size is a critical factor to evaluate hardware cost, we recited here the total buffer size and the size of each virtual-channel for each type of router for reference. Note that for the conventional unidirectional NoCs in a mesh topology, a 5×5 crossbar was adequate to support the 5-in 5-out data transmission. Nevertheless, in BiNoC architectures, a bigger 10×10 crossbar was needed to provide flexible data transmission in the 10-in/out bidirectional channels.

To evaluate the performance of each type of routers in a fairness way, each selected router architectures was performed under its maximum supported frequency, which was calculated by Synopsys Design Compiler under United Microelectronics Corporation (UMC) 90 nm technology, as illustrated in Table 6.1, and normalized to the T-NoC_4VC(32) architecture for easier comparison. We can find that BiNoC_WH(32) had the shortest logic delay among all router models since the allocation module needed in a wormhole flow-control based router was much less than in a virtual-channel flow-control based router. In other words, the size of the allocation module had a great impact on the critical path of routers. In addition, the logic delay also grew with thc number of virtual-channels implemented in the router due to the latency expensed in the VA stage. The timing difference between T-NoC_4VC(32) and BiNoC_4VC(32) with the same number of virtual-channels was ascribed to the implementation overhead in achieving bidirectional data transmission. This latency overhead was mainly distributed in the SA stage and the ST stage because of the doubled amount of channel requests processed by the output arbiter and the larger sized crossbar.

For fair evaluation, the latency results in the following sections were calculated in unit normalized to the cycle time of the T-NoC_4VC(32) architecture. For example, one simulation cycle for BiNoC_WH(32) was converted to 0.81 *normalized cycle* in comparison.

6.5.2 Synthetic Traffic Analysis

In synthetic traffic analysis, three types of traffic patterns were run: uniform, transpose, and regional. The detailed features of these three types of traffic patterns are described in Appendix A. Each type of test performed generates packets with a varying injection rate for 25,000 cycles.

6.5.2.1 Comparison between BiNoC and Conventional NoC

Figure 6.22 illustrates the latency versus injection rate results obtained by running XY routing under Fig. 6.22a uniform,c regional, e transpose traffics, and running Odd–Even routing under Fig. 6.22b uniform, d regional, f transpose traffics.

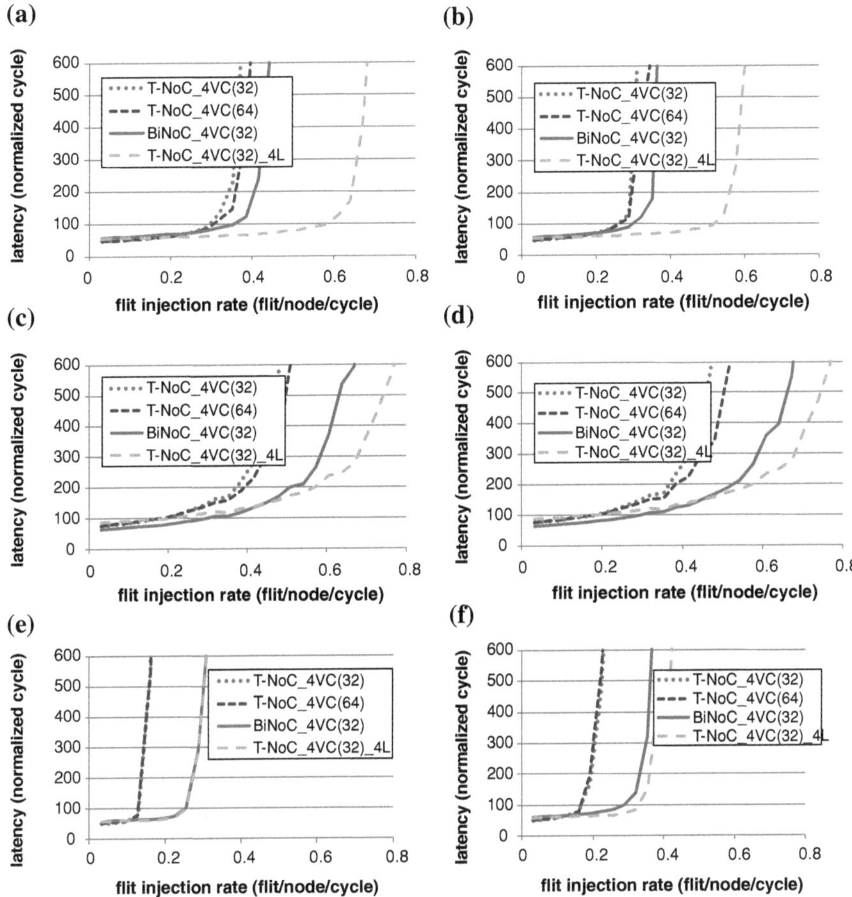

Fig. 6.22 Latency versus injection rate results of BiNoC and T-NoC

We can find that our proposed BiNoC_4VC(32) had a lower latency than the T-NoC_4VC(32) under both XY and Odd–Even routing algorithms in these three traffic patterns even if compared with T-NoC_4VC(64) which has a double buffer size. Comparing results in uniform traffic with regional traffic, we can find that both T-NoC and BiNoC architectures performed better in terms of latency versus injection rate under the regional traffic case, since the uniform pattern sends packets randomly and resulted in more traffic contention. However, in the regional traffic case, more packets transmitted to the destination near around reduced the traffic contention on a network, which lowered the direction switching frequency of a bidirectional channel in a BiNoC and gathered more latency improvement than a T-NoC.

In the transpose traffic case, such regular traffic patterns will make many hardwired unidirectional channels connecting neighboring routers unable to be used. Therefore, the latency results of a T-NoC were restricted by its bandwidth utilization capability as illustrated in Fig. 6.22e, f. Enlarging the buffer size of a T-NoC_4VC(32) to a T-NoC_4VC(64) cannot improve the latency results even using adaptive routing algorithm such as Odd–Even. However, for the BiNoC architecture, more packets can be transmitted via the two bidirectional channels, which will break the bottleneck caused by unidirectional channels, and achieve better latency result than a T-NoC.

As to the comparison between a BiNoC_4VC(32) with two bidirectional links and a T-NoC_4VC(32)_4L with four unidirectional links, T-NoC_4VC(32)_4L performed better than BiNoC_4VC(32) in uniform and regional traffics, since it can afford double bandwidth for inter-router data transmission. However, the latency difference for these two architectures in regional traffic was much closer than in uniform traffic since most of data transmissions were within three hops away. Furthermore, the latency results in transpose traffic with XY routing applied was almost the same as illustrated in Fig. 6.22e since regular traffic patterns made about half amount of unidirectional links unable to use.

6.5.2.2 Performance Under Equal Buffer Depth

Figure 6.23 shows the comparison of wormhole flow-control with virtual-channel flow-control by latency versus injection rate. Simulation results were obtained by running XY routing under Fig 6.23a uniform, c regional, and e transpose traffics, and running Odd–Even routing under Fig 6.23b uniform, d regional, f transpose traffics.

From the above figure, we can find that the effects of wormhole flow-control and virtual-channel flow-control toward our BiNoC architecture in terms of latency result versus various flit injection rates under the same buffer depth. As illustrated in Fig. 6.23a–d, BiNoC_2VC(32) with virtual-channels performed much better than BiNoC_WH(32) running under wormhole flow-control using an equal input buffer size of 32 (the two input buffers, each having a size of 16, can be implemented as two independent FIFOs for BiNoC_WH or two virtual-channels for BiNoC_2VC) in each direction. This is because the virtual-channel diminishes the blocking probability caused by wormhole flow-control. However, BiNoC_WH(32) performed better than other virtual-channel flow-control architectures in transpose traffic with XY routing applied as illustrated in Fig. 6.23e, because virtual-channel flow-control and adaptive routing raised the physical channel utilization probability in trade of using a more complex arbitration technique and one more stage of router pipeline, which was not necessary for regular traffic such as transpose when packets can always flow smoothly.

Regarding the influence of the increasing number of virtual-channels, BiNoC_2VC(32), BiNoC_3VC(48) and BiNoC_4VC(64) represent three BiNoC architectures using different numbers of virtual-channels while each virtual-channel had an equal depth of 16 flits. Figure 6.24 illustrates comparison of

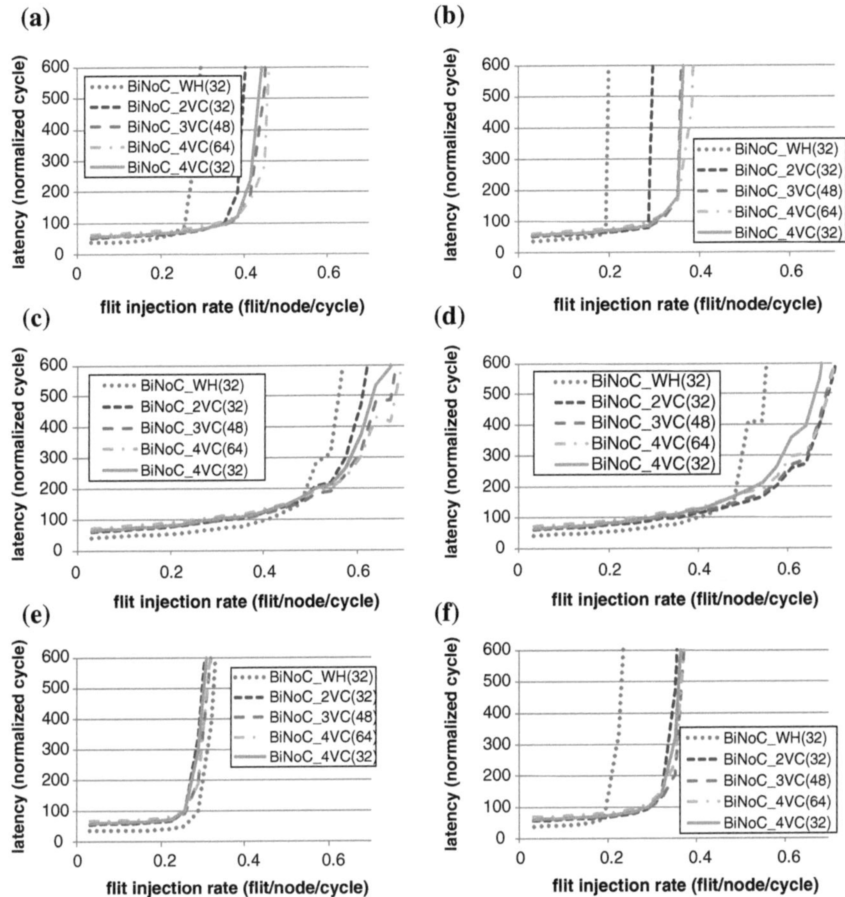

Fig. 6.23 Comparison of flow-control under equal buffer depth

wormhole flow-control with virtual-channel flow-control by latency versus injection rate. Simulation results were obtained by running XY routing under Fig 6.24a uniform, c regional, e transpose traffics, and running Odd–Even routing under Fig 6.24b uniform, d regional, f transpose traffics. These configurations of virtual-channel numbers under three traffic patterns demonstrated the network performance improvement by allowing the active packets to have a chance to pass the blocked packets. Note that the latency improvement by increasing the number of virtual-channels from 2 to 3 was much better than increasing the number of virtual-channels from 3 to 4. Besides, BiNoC_4VC(32) generated better latency results than BiNoC_2VC(32) but worse than BiNoC_3VC(48) in most of the results. Thus, we can improve performance by increasing the number of virtual-channels or extending the length of each existing virtual-channel under different implementation overhead considerations.

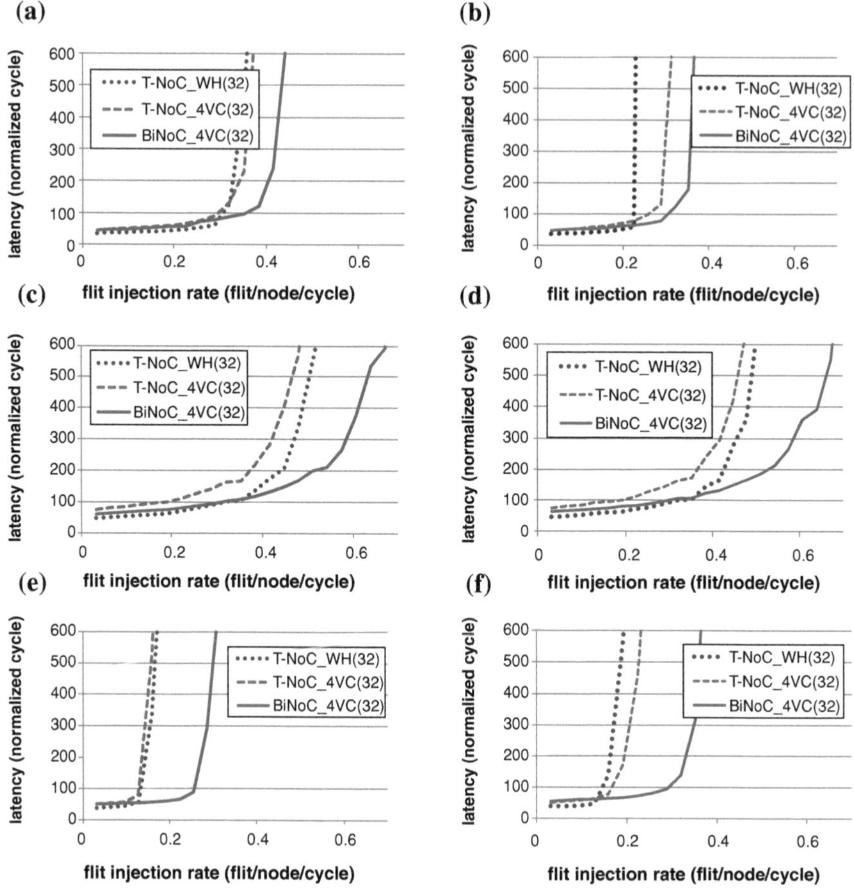

Fig. 6.24 Comparison of flow-control under equal buffer size

6.5.2.3 Performance Under Equal Buffer Size

Regarding the comparison of typical NoC with wormhole flow-control, we can find that even the T-NoC_WH architectures can be implemented in a three-stage pipeline with higher frequency, the BiNoC architectures still performed better than the typical NoC despite the type of flow-control and routing methods used as illustrated in Fig. 6.24. However, in regional and transpose traffics, T-NoC_WH architecture performed better than T-NoC_4VC as illustrated in Fig. 6.24c–e. This is because the regional and transpose traffics cause less contention between packets compared to uniform traffic.

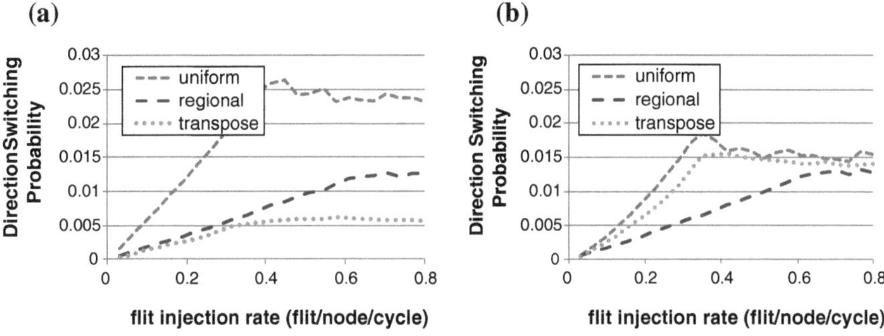

Fig. 6.25 Switching probability of each bidirectional channel

6.5.2.4 Bidirectional Channel Switching Probability

By observation above, the diversity of traffic also affects the performance of a bidirectional channel because of the times needed to configure the channel direction. Figure 6.25a, b show the switching probabilities of bidirectional channels BiNoC_4VC(32) obtained by running the above three kinds of synthetic patterns under XY and Odd–Even routing respectively.

The uniform traffic sends packets randomly and causes the bidirectional channel to switch direction frequently. Thus, more configuration time was wasted to operate the channel reconfiguration by switching its direction than other two traffic types. On the other hand, lower switching probabilities were observed in the regional and transpose patterns because of their lower traffic diversity. However, even in the uniform traffic, channel-direction switching probability was not higher than 2.5% in both routing algorithms since our bidirectional channel control logic rendered higher priority to the packets traveling in the same direction to decrease the channel-direction switching probability.

6.5.2.5 Bandwidth Utilization Analysis

Figure 6.26 illustrates the bandwidth utilization versus injection rate results obtained by running XY routing under Fig 6.26a uniform, c regional, e transpose traffics, and running Odd–Even routing under Fig 6.26b uniform, d regional, f transpose traffics. We can find that the BiNoC architecture always had the best saturation performance because of its flexibility.

In observation of regular traffic patterns such as regional traffic and transpose traffic, the bandwidth utilization improvement percentages were much better than uniform traffic since more idle channels can be dynamically reconfigured to transmit data. Moreover, the bandwidth utilization results show that BiNoC_4VC(32) under these three synthetic traffic patterns performed even better than T-NoC_4VC(64) which had a doubled buffer size. T-NoC_4VC(32)_4L was

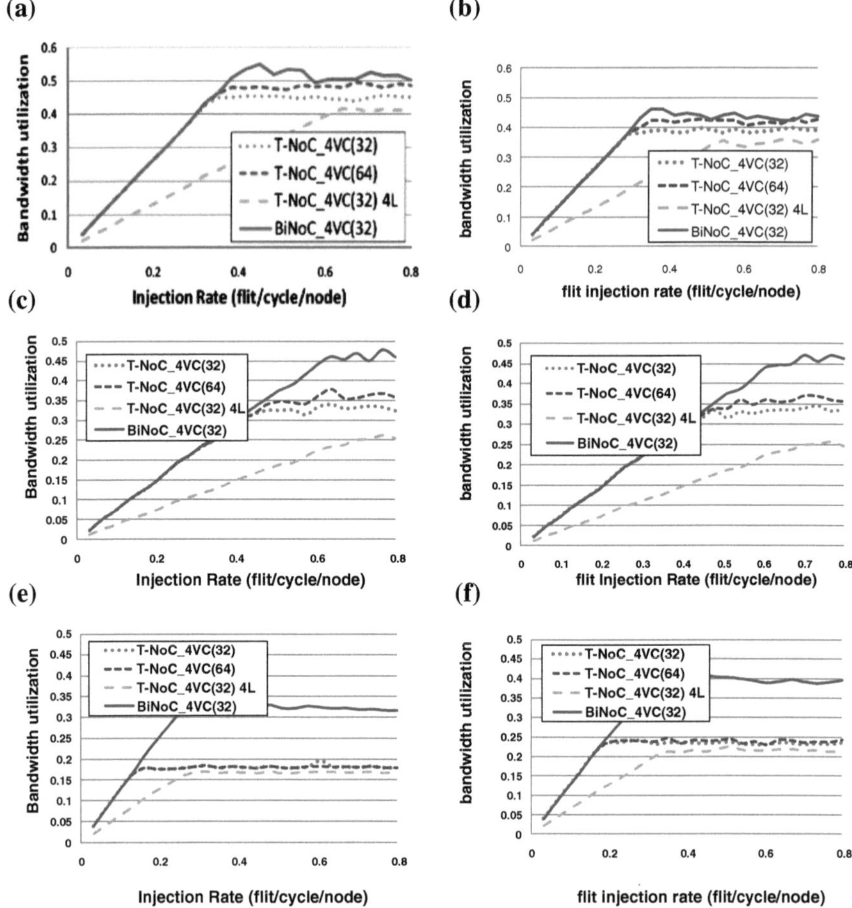

Fig. 6.26 Bandwidth utilization analysis

always the worst one since additional wires cannot always be used but will cause additional physical overhead.

6.5.2.6 Buffer Size Analysis

To understand the effects caused by buffer size, we used regional traffic with Odd–Even routing to run simulation, because it is more realistic compared to the uniform and transpose traffics. Several router architectures with various depths in each buffer queue were adopted in our experiments. Figure 6.27a, b respectively illustrate latency versus buffer-size analysis over regional traffic patterns while different buffers (two-virtual-channel vs four-virtual-channel) in each direction of the routers were implemented to compare.

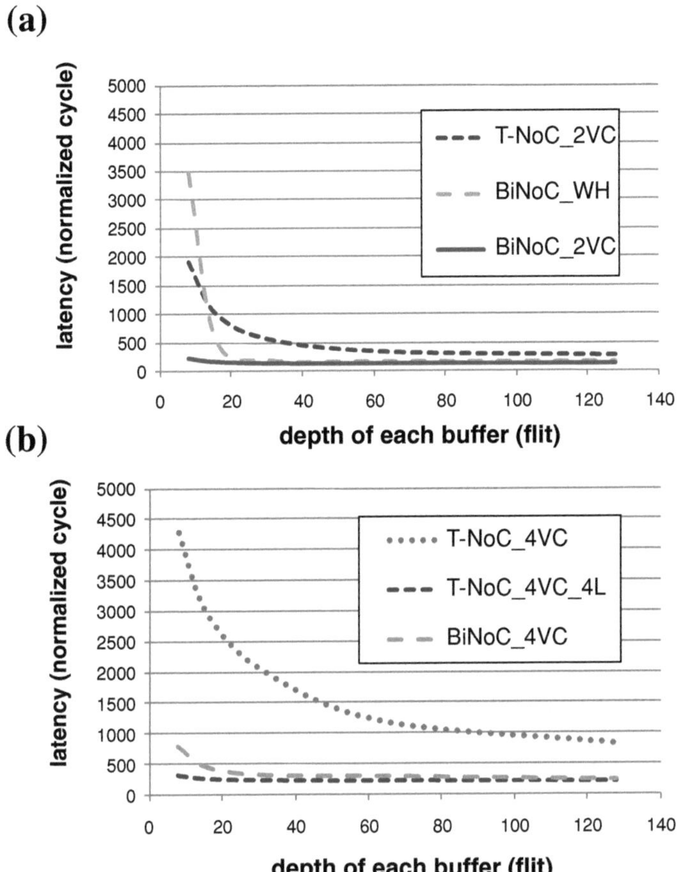

Fig. 6.27 Latency versus buffer size analysis

The latency diagrams were separated according to the number of total buffer queues in each direction of the router architectures for fair comparison. Figure 6.27a shows the results of simulation on T-NoC_2VC, BiNoC_WH, and BiNoC_2VC, which contained two buffer queues in each direction, under different queue depths at a flit injection rate of 0.512. We can observe that as buffer size decreased, the latencies of BiNoC_2VC were not increased dramatically as the other two architectures did. In other words, our proposed virtual-channel flow-control based BiNoC architecture can achieve better buffer utilization efficiency thus also increase the physical channel utilization.

Another quite valuable phenomenon appeared in these simulation results was that NoC performance cannot keep improving while the buffer size is increased. For instance, increasing the depth of each queue from 8 to 32 flits can achieve a great deal of performance improvement in a T-NoC_2VC, while there was a slight

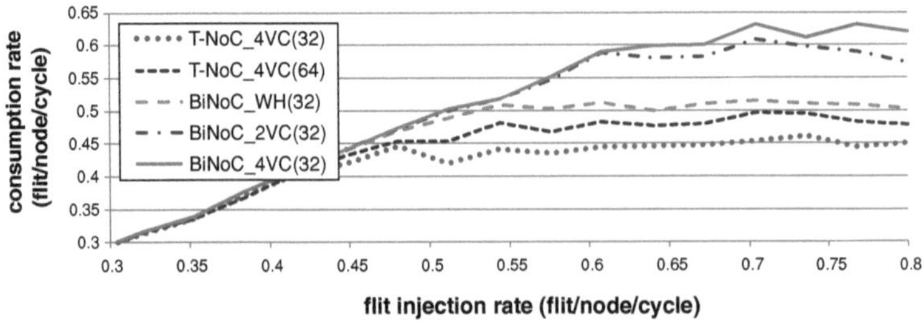

Fig. 6.28 Consumption rate versus injection rate

latency reduction if the queue size increased from 32 to 128 flits. However, we find that our BiNoC, if limited with fewer buffer area cost, still achieved a better performance than a typical NoC did. Furthermore, virtual-channel flow-control performed better than wormhole flow-control on our BiNoC architecture while the buffer size was small. The latency comparison of T-NoC_4VC and BiNoC_4VC at a flit injection rate of 0.704 also exhibited significant performance superiority in the BiNoC router with dynamic reconfigurable bidirectional channels as illustrated in Fig. 6.27b.

6.5.2.7 Traffic Consumption Rate Analysis

Figure 6.28 presents the performance results of consumption rate versus injection rate obtained by using the regional traffics to run simulation since it is more realistic compared to the other two. Observation was obtained at low injection rate, which matched with the consumption rates of an NoC. But after reaching a definite injection rate, the flits were no longer being consumed at the same rate as theywere input to the NoC.

The phenomenon that the NoC injection rate is increasing faster than their consumption rate is analogous to the fundamental traffic model that reaches its critical density point [22]. Note that our BiNoC architecture can exhibit higher critical densities than the T-NoC_4VC architecture, even when we use less buffer size. Under the same total buffer size restriction, applying virtual-channel flow-control to BiNoC architectures performed better in consumption rate than applying wormhole flow-control.

6.5.3 Experiments with Real Applications

According to the experimental results above, the channel configuration time had important impact on transmission latency. However, data transmission in real cases was much more regular than synthetic patterns. Therefore, we used E3S benchmarks from the Embedded Microprocessor Benchmark Consortium (EEMBC) [23] to demonstrate the improvements of our BiNoC architectures in this section. The three tests used are auto-indust, consumer, and telecom which were run on a 5 ×5, a 4 × 4, and a 6 × 6 NoC architectures, respectively. The detailed behavior of E3S traffic patterns is explained in Appendix A.

Figure 6.29a–c respectively illustrate the latency versus injection rate results obtained by running E3S benchmarks of auto-indust, consumer, and telecom under Odd–Even routing. As illustrated in Fig. 6.29, BiNoC_4VC(32) consistently provided better latency results than T-NoC_4VC(32) and T-NoC_4VC(64) among the three benchmark applications. This is because the proposed self-reconfigurable bidirectional channel had more flexibility for data transmission. In real traffic, we can find that T-NoC_4VC(32) had almost the same latency with respect to the T-NoC_4VC(64) because of the regularity characteristic in real traffics. In other words, the fixed unidirectional channel of a T-NoC restricts its own performance, hence increasing buffer size from 32 to 64 in each direction cannot help much while using the same task mapping algorithm. As to the comparison of BiNoC_WH(32) and BiNoC_4VC(32) which had the same buffer size with different flow-control mechanisms, the effects of increasing the number of virtual-channels only resulted in minor latency improvement. Therefore, this is another evidence showing that the most serious bottleneck of latency was bounded by physical channel utilization as we suspected but not by the buffer size or number of virtual-channels, especially under a well-managed traffic mapping.

Regarding T-NoC_4VC(32)_4L, its performance results in E3S benchmark were almost the same with BiNoC_4VC(32). Therefore, we know that as long as the tasks of real applications are mapped properly in advance, BiNoC architecture with two bidirectional links can also generate the same good latency results as the typical NoC with 4 unidirectional links. In other words, if we can map the tasks of an application into appropriate locations in advance by a software mapper, BiNoC can also provide good latency results without extra hardware overhead.

6.5.4 *Implementation Details in Terms of Area and Power*

The router architectures were described in HDL and synthesized in UMC CMOS 90 nm technology under typical operating conditions, 1.0 V and 25°C. Furthermore, as well as the latency results observed in previous section, the area and power of selected router architectures were also measured at their maximum supported frequencies as illustrated in Table 6.1.

6.5.4.1 Area Measurement

Table 6.2 lists the hardware implementation area cost of six NoC router architectures in terms of logic gate count and percentage calculated by *Synopsys Design*

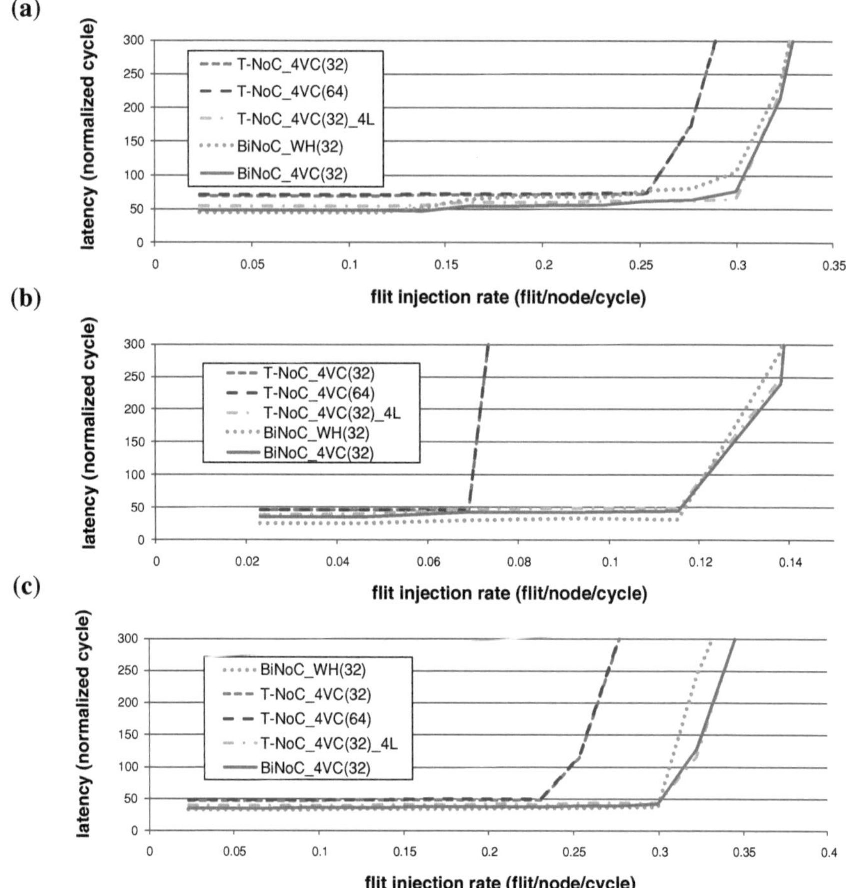

Fig. 6.29 Performance evaluation of E3S benchmarks. **a** Auto-indust. **b** Consumer. **c** Telecom

Compiler. For the sake of fairness, the first five NoC routers in this comparison were equipped with the same amount of buffer size (32 flits) in each direction for wormhole or virtual-channel flow-control.

In observation, the input buffer stages and virtual-channel allocation occupied most of the router area. As the experimental results in Zhang's work [24] based on the router architecture proposed in [16], VA logic occupied around 25% of the router area in a 5-in, 5-out, 4 virtual-channels, 4 32bit-flits per virtual-channel, XY routing router. Since the number of available virtual-channels in a router rules the size of the VA stage, a BiNoC_4VC architecture required similar amount of gate count to implement this stage as a T-NoC_4VC architecture. Therefore, there is no additional area overhead at this stage.

Regarding the input buffer stage, each input buffer queue was implemented using shift registers with a multiplexer-based input indicated by a *tail* pointer.

Table 6.2 Area breakdown of different NoC architectures

Component	Area (gate count) ∣ (%)					
Buf./dir.	T-NoC_WH(32) 32 flits × 1	BiNoC_WH(32) 16 flits × 2	T-NoC_4VC(32) 8 flits × 4	BiNoC_4VC(32) 8 flits × 4	T-NoC_4VC(32)_4L 8 flits × 4	T-NoC_4VC(64) 16 flits × 4
Input buf. + buf. ctrl	39,699 ∣ 94.50	38,668 ∣ 79.46	36,754 ∣ 70.87	37,413 ∣ 63.75	37,542 ∣ 65.26	66,067 ∣ 79.51
Routing computation	186 ∣ 0.44	372 ∣ 0.76	629 ∣ 1.21	667 ∣ 1.13	664 ∣ 1.15	629 ∣ 0.76
VC allocation	–	–	12,103 ∣ 23.33	12,321 ∣ 20.99	12,662 ∣ 22.01	14,024 ∣ 16.88
Switch allocation	664 ∣ 1.58	3,082 ∣ 6.33	1,006 ∣ 1.93	2,248 ∣ 3.83	2,248 ∣ 3.90	1,006 ∣ 1.21
Switch traversal	1,463 ∣ 3.48	4,694 ∣ 9.64	1,365 ∣ 2.63	4,404 ∣ 7.50	4,404 ∣ 7.65	1,365 ∣ 1.64
Bidir. ch. ctrl	–	1,850 ∣ 3.80	–	1,628 ∣ 2.77	–	–
Total	42,011 ∣ 100.00	48,666 ∣ 100.00	51,857 ∣ 100.00	58,681 ∣ 100.00	57,520 ∣ 100.00	83,091 ∣ 100.00
Normalized area	0.81	0.93	1.00	1.13	1.10	1.60

While the flit stored at the header of the buffer queue is being read, the subsequent flits in the register-based FIFO buffer are shifted one position towards the header. A virtual-channel flow-control based router needs additional channel selection logics for this stage. Thus, even we implemented router architectures under the same buffer size, different flow-control mechanisms with various buffer depths resulted in different sizes of buffer control logic at this stage.

Comparing conventional T-NoC with BiNoC under the same flow-control mechanism with the same total buffer size, as shown in Table 6.2, we find that the major area overhead of BiNoC was caused by the SA and ST stages. Regarding SA, the doubled numbers of channel requests and usable channels of a BiNoC architecture increased the logic complexity. Regarding ST, since BiNoC needs a 10×10 crossbar to enable bidirectional transmission, the area of BiNoC crossbar was around three times bigger than the typical NoC one. However, the size of ST varied to the architecture of a T-NoC or BiNoC router while the depth of input buffer can be adjusted depending on need. According to our implementation results, the ST stages of BiNoC_WH(32) and BiNoC_4VC(32), that had a 32-flits buffer in each direction, occupied 9.64 and 7.50% of their respective total gate count. Nevertheless, the percentage of ST stage in BiNoC increased as the buffer size decreased.

6.5.4.2 Power Measurement

Table 6.3 lists the power breakdown across the major component of various router architectures (each running at its maximum supported frequency) which were calculated by *Synopsys PrimePower* under a regional traffic of flit injection rate 0.322, where the latency results of all selected router architectures still performed normally before reaching saturation throughput [25]. Packets of 16-flits with a 50% switching activity factor of random payload were used in this experiment. Power measurements were performed for 10,000 cycles where the initial 1,000 cycles were used as the warm-up time. Power consumption of each router running at 500 MHz was also provided in Table 6.3.

Observing the power consumption in Table 6.3 at an operating frequency of 500 MHz, all the selected routers, which were equipped with a 32-flits input buffer in each direction, expensed similar amount of power since the data-path components dominated over the control elements in various routers [26]. However, if we compare power consumption of each router running at its maximum supported frequency consistently with the way we compared the latency results, BiNoC always expensed less power than T-NoC in both wormhole and virtual-channel flow-control mechanisms.

Comparing T-NoC_WH(32) with BiNoC_WH(32) under its maximum supported frequency, even BiNoC had a higher power consumption at the SA and ST stages, the total power expenditure for T-NoC_WH(32) was still larger. One reason is that the 32-flits buffers in a BiNoC_WH(32) architecture were implemented in two independent 16-flit-long register-based FIFOs, thus these 32 flits of

Table 6.3 Power breakdown of different NoC architectures

Component	Power (mW) I (%)					
	T-NoC_WH(32)	BiNoC_WH(32)	T-NoC_4VC(32)	BiNoC_4VC(32)	T-NoC_4VC(32)_4L	T-NoC_4VC(64)
Freq. (MHz)	500	500	500	500	500	500
Router (mW)	30.29	29.89	32.31	35.12	35.29	51.84
Max. freq. (MHz)	1166	921	746	627	632	746
Input buf. + buf. ctrl	70.43 I 95.43	46.23 I 86.91	38.79 I 80.48	32.02 I 72.71	33.74 I 75.64	69.89 I 88.08
Routing computation	0.30 I 0.41	0.56 I 1.05	0.52 I 1.08	0.46 I 1.05	0.47 I 1.05	0.53 I 0.66
VC allocation	–	–	6.22 I 12.91	5.79 I 13.14	5.86 I 13.13	6.24 I 7.86
Switch allocation	0.83 I 1.13	2.69 I 5.06	0.90 I 1.87	1.91 I 4.33	1.61 I 3.60	0.90 I 1.13
Switch traversal	2.25 I 3.05	2.93 I 5.51	1.78 I 3.68	2.99 I 6.78	2.94 I 6.59	1.79 I 2.25
Bidir. ch. ctrl	–	0.80 I 1.50	–	0.87 I 1.98	–	–
Total	73.81 I 100.00	53.21 I 100.00	48.21 I 100.00	44.04 I 100.00	44.60 I 100.00	79.34 I 100.00
Normalized area	1.53	1.10	1.00	0.91	0.92	1.64

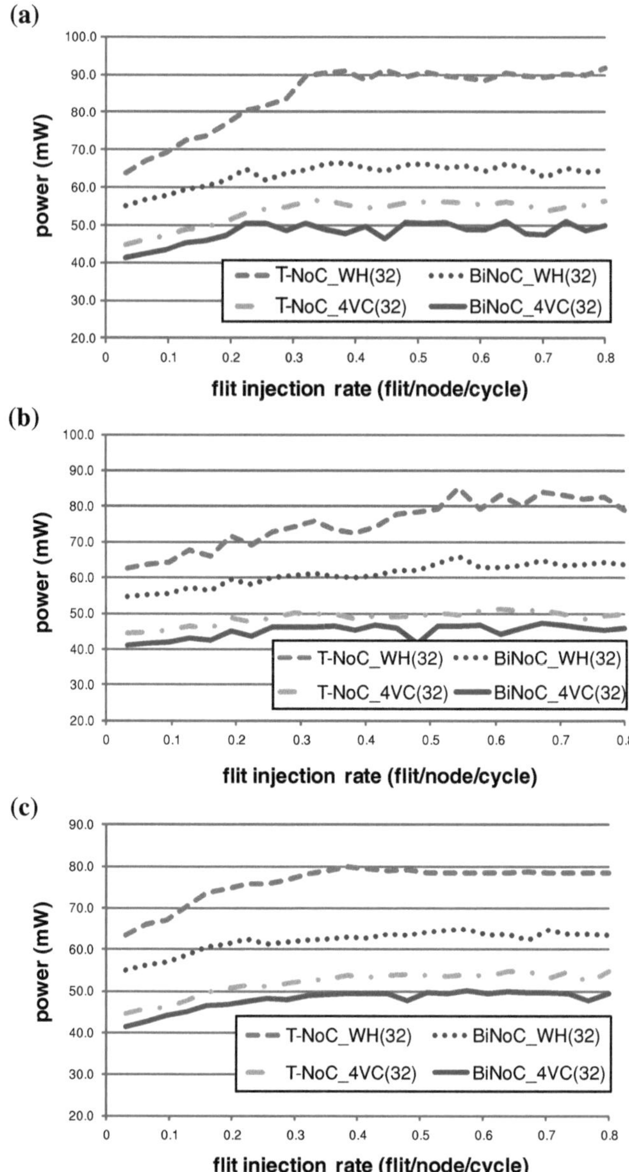

Fig. 6.30 Power versus injection rate

registers were not necessarily triggered at the same time as a T-NoC_WH(32) architecture and lowered the dynamic power. Another important reason is that the maximum supported frequencies of BiNoC routers were lower than T-NoC routers, and the power consumption was proportional to the operating frequency.

The consistent results as shown in the virtual-channel flow-control based router, BiNoC_4VC(32), with its overhead gate count in SA, ST, and channel-direction reconfiguration logic, still consumed 9% less power than T-NoC_4VC(32).

Figure 6.30a–c respectively illustrate the power results for various injection rates by running uniform, regional, and transpose traffic patterns under maximum supported frequency of each router. The power consumption for each of the selected routers was increased with traffic load, which resulted in more data-path computation.

However, when the congestion happened, the power consumption did not increase with the congestion seriously since the stuck data stored in the clock-gated register did not need to be flipped; thus, the amount of data-path computation, which dominates the power expenditure, did not increase seriously. In observation of Fig. 6.30, BiNoC always had better power consumption than T-NoC router in both flow-control mechanisms because the maximum supported frequencies for BiNoC architectures were lower than T-NoC architectures. Therefore, according to the latency results gathered by each router running at its maximum supported frequency and calculated by *normalized cycles* in previous sections, we can conclude that even the maximum supported frequencies of BiNoC architectures are lower than T-NoC architectures, BiNoC can perform better in latency while requiring less power.

6.5.5 Implementation Overhead

In order to enable bidirectional data transmission, several modifications should be made across router stages. In the following sections, we will describe in detail the implementation overhead of these modifications to justify these valuable changes.

6.5.5.1 Overhead at Bidirectional Channel Control Logic

Regarding the cost of bidirectional channel control logic in BiNoC, the control logic implemented in FSM is much smaller than other function blocks of the router, because it only occupied less than 4 and 2% of the entire router in terms of area and power respectively, as illustrated in Tables 6.2 and 6.3.

6.5.5.2 Overhead at ST Stage

Regarding the crossbar (ST stage) overhead comparison between BiNoC and T-NoC, it is apparent that the area of a 10×10 crossbar could be roughly four times higher than a 5×5 crossbar. Normally, the number was slightly lower than four because of the optimization algorithm implemented in the logic synthesis CAD tools. Therefore, in Table 6.2, we can observe that the gate count number of

Table 6.4 Area breakdown of T-NoC_WH

Component	T-NoC_WH T-NoC_WH Architectures area (gate count) ǀ (%)			
Buf./Dir.	T-NoC_WH(4) 4 flits × 1	T-NoC_WH(8) 8 flits × 1	T-NoC_WH(16) 16 flits × 1	T-NoC_WH(32) 32 flits × 1
Input buf. + buf. ctrl	5,218 ǀ 69.34	9,665 ǀ 80.75	1,9544 ǀ 89.43	39,699 ǀ 94.50
Routing computation	186 ǀ 2.47	186 ǀ 1.55	186 ǀ 0.85	186 ǀ 0.44
VC allocation	–	–	–	–
Switch allocation	657 ǀ 8.73	656 ǀ 5.48	660 ǀ 3.02	664 ǀ 1.58
Switch traversal	1,463 ǀ 19.45	1,463 ǀ12.22	1,465 ǀ 6.70	1,463 ǀ 3.48
Bidir. ch. ctrl	–	–	–	–
Total	7,524 ǀ 100.00	11,970 ǀ 100.00	21,855 ǀ 100.00	42,011 ǀ 100.00

the ST stage in a BiNoC_4VC(32) was 3.22 (4404/1365) times higher than in a T-NoC_4VC(32) router.

However, to compare the power consumption between a 10 × 10 and a 5 × 5 crossbar switches, besides the size difference, the actual data throughput across the crossbar should be a much more important factor since the dynamic power frequently dominated the power expenditure. In a well-designed hardware, when there is no data traversing across the crossbar, the circuit should be idle and hence consumes less power. Thus, in NoC, the power consumption was dependent on the data injection rate as shown in Fig. 6.30.

Moreover, although a 10 × 10 crossbar switch should provide twice available bandwidth than a 5 × 5 crossbar switch, due to traffic congestion, the observed actual maximum throughput improvement offered by a 10 × 10 crossbar switch was between 1.2 and 1.8 times (depends on the traffic type) over that of a 5 × 5 crossbar switch as shown in Fig. 6.26. Therefore, depending on the flit injection rate and actual data throughput, the power consumption in ST stage for a BiNoC router was less than 4 times over the power consumption in ST stage for a T-NoC router.

6.5.5.3 Overhead at Input Buffer Stage

According to Tables 6.2 and 6.3, we find that our BiNoC_4VC(32) had only about 13% area (gate count) overhead over the T-NoC_4VC(32) because of its doubled crossbar and SA logic. However, the latency results as illustrated in Fig. 6.27 showed that a BiNoC_4VC(32) can even outperform a T-NoC_4VC(64) (which had respectively 60% area and 64% power over a T-NoC_4VC(32) because of its double-sized buffers).

In addition, as illustrated in Tables 6.4, 6.5, 6.6, and 6.7, we can observe that as buffer size increased, the relative (percentage) area overhead due to crossbar switch decreased. Thus, we can conclude that, since the buffer memory dominated the major cost of router, we achieved the goal of hardware saving by adopting a smaller buffer size in a BiNoC than a T-NoC router while getting better performance.

Table 6.5 Area breakdown of T-NoC_4VC

Component	T-NoC_4VC Architectures area (gate count) | (%)		
Buf. depth	T-NoC_4VC(16) 4 flits × 4	T-NoC_4VC(32) 8 flits × 4	T-NoC_4VC(64) 16 flits × 4
Input buf. + buf. ctrl	21,746 | 56.30	36,754 | 70.87	66,067 | 79.51
Routing computation	629 | 1.63	629 | 1.17	629 | 0.76
VC allocation	13,882 | 35.94	12,103| 23.33	14,024 | 16.88
Switch allocation	1,005 | 2.60	1,006 | 1.93	1,006 | 1.21
Switch traversal	1,365 | 3.53	1,365 | 2.63	1,365 | 1.64
Bidir. ch. ctrl	–	–	–
Total	38,627 | 100.00	51,857| 100.00	83,091 | 100.00

Table 6.6 Area breakdown of BiNoC_WH

Component	BiNoC_WH Architectures Area (gate count) | (%)		
Buf. depth	BiNoC_WH(8) 4 flits × 2	BiNoC_WH(16) 8 flits × 2	BiNoC_WH(32) 16 flits × 2
Input buf. + buf. ctrl	8,829 | 46.90	19,123 | 65.66	38,668 | 79.46
Routing computation	372 | 1.98	372 | 1.28	372 | 0.76
VC allocation	–	–	–
Switch allocation	3,081 | 16.37	3,084 | 10.59	3,082 | 6.33
Switch traversal	4,694 | 24.93	4,694 | 16.12	4,694 | 9.64
Bidir. ch. ctrl	1,850 | 9.83	1,850 | 6.35	1,850 | 3.80
Total	18,826 | 100.00	29,123 | 100.00	48,666 | 100.00

Table 6.7 Area breakdown of BiNoC_4VC

Component	BiNoC_4VC Architectures Area (gate count) | (%)		
Buf. depth	BiNoC_4VC(16) 4 flits × 4	BiNoC_4VC(32) 8 flits × 4	BiNoC_4VC(64) 16 flits × 4
Input buf. + buf. ctrl	18,724 | 46.85	37,413 | 63.75	66,718 | 75.90
Routing computation	667 | 1.66	667 | 1.13	667 | 0.75
VC allocation	12,297 | 30.77	12,321 | 20.99	12,246 | 13.93
Switch allocation	2,243 | 5.61	2,248 | 3.83	2,241 | 2.54
Switch traversal	4,404 | 11.02	4,404 | 7.50	4,404 | 5.01
Bidir. ch. ctrl	1,626 | 4.06	1,628 | 2.77	1,626 | 1.84
Total	39,960 | 100.00	58,681 | 100.00	87,902 | 100.00

The impacts of increasing buffer size (Input Buffer and Buffer Control Logic) to the relative power consumption by crossbar switch (ST) were also investigated. Based on the implementation of T-NoC_WH, T-NoC_4VC, BiNoC_WH and BiNoC_4VC, the results were illustrated in Tables 6.8, 6.9, 6.10, and 6.11 respectively. It is quite clear that the percentage power consumption of crossbar switch (ST) decreased as the buffer size increased.

Table 6.8 Power breakdown of T-NoC_WH

Component	T-NoC_WH Architectures Power [mW] ǀ (%)			
	T-NoC_WH(4)	T-NoC_WH(8)	T-NoC_WH(16)	T-NoC_WH(32)
Input buf. + buf. ctrl	8.04 ǀ 72.36	17.90 ǀ 84.15	37.54 ǀ 91.35	70.43 ǀ 95.43
Routing computation	0.24 ǀ 2.25	0.29 ǀ 1.39	0.31 ǀ 0.78	0.30 ǀ 0.41
VC allocation	–	–	–	–
Switch allocation	0.84 ǀ 7.60	0.83 ǀ 3.94	0.84 ǀ 2.05	0.83 ǀ 1.13
Switch traversal	1.97 ǀ 17.80	2.23 ǀ 10.51	2.39 ǀ 5.82	2.25 ǀ 3.05
Bidir. ch. ctrl	–	–	–	–
Total	11.11 ǀ 100.00	21.28 ǀ 100.00	41.09 ǀ 100.00	73.81 ǀ 100.00

Table 6.9 Power breakdown of T-NoC_4VC

Component	T-NoC_4VC Architectures Power (mW) ǀ (%)		
Buf. depth	T-NoC_4VC(16) 4 flits	T-NoC_4VC(32) 8 flits	T-NoC_4VC(64) 16 flits
Input buf. + buf. ctrl	21.12 ǀ 69.97	38.79 ǀ 80.48	67.88 ǀ 87.78
Routing computation	0.49 ǀ 1.64	0.52 ǀ 1.08	0.52 ǀ 0.68
VC allocation	6.19 ǀ 20.51	6.22 ǀ 12.91	6.23 ǀ 8.06
Switch allocation	0.86 ǀ 2.87	0.89 ǀ 1.87	0.90 ǀ 1.17
Switch traversal	1.51 ǀ 5.02	1.77 ǀ 3.68	1.79 ǀ 2.31
Bidir. ch. ctrl	–	–	–
Total	30.19 ǀ 100.00	48.21 ǀ 100.00	77.33 ǀ 100.00

Table 6.10 Power breakdown of BiNoC_WH

Component	BiNoC_WH Architectures power (mW) ǀ (%)		
Buf. depth	BiNoC_WH(8) 4 flits	BiNoC_WH(16) 8 flits	BiNoC_WH(32) 16 flits
Input buf. + buf. ctrl	11.52 ǀ 63.11	23.39 ǀ 77.40	46.23 ǀ 86.91
Routing computation	0.34 ǀ 1.92	0.35 ǀ 1.18	0.56 ǀ 1.05
VC allocation	–	–	–
Switch allocation	2.68 ǀ 14.73	2.68 ǀ 8.90	2.69 ǀ 5.06
Switch traversal	2.90 ǀ 15.93	2.96 ǀ 9.81	2.93 ǀ 5.51
Bidir. ch. ctrl	0.78 ǀ 4.32	0.82 ǀ 2.72	0.80 ǀ 1.50
Total	18.26 ǀ 100.00	30.22 ǀ 100.00	53.21 ǀ 100.00

6.5.5.4 Overhead at Interconnection Wires

Regarding the typical NoC with 4 inter-router unidirectional links, we can find that the area of T-NoC_4VC(32)_4L was almost the same with BiNoC_4VC(32) since the size of crossbar and the number of virtual-channels were all the same. The only difference was the bidirectional channel control logic in BiNoC. However, the

Table 6.11 Power breakdown of BiNoC_4VC

Component	BiNoC_4VC Architectures power (mW) I (%)		
Buf. Depth	BiNoC_4VC(16) 4 flits	BiNoC_4VC(32) 8 flits	BiNoC_4VC(64) 16 flits
Input buf. + buf. ctrl	16.95 I 60.68	32.02 I 72.71	53.42 I 82.46
Routing computation	0.43 I 1.55	0.46 I 1.05	0.44 I 0.68
VC allocation	5.78 I 20.69	5.79 I 13.14	5.85 I 9.04
Switch allocation	1.73 I 6.19	1.91 I 4.33	1.79 I 2.77
Switch traversal	2.40 I 8.60	2.99 I 6.78	2.58 I 3.99
Bidir. ch. ctrl	0.63 I 2.28	0.87 I 1.98	0.68 I 1.06
Total	27.94 I 100.00	44.04 I 100.00	64.78 I 100.00

power consumption of T-NoC_4VC(32)_4L was slightly higher than BiNoC_4VC(32) since it had better latency results and can run at higher frequency. However, even the overhead of the T-NoC_4VC(32)_4L router itself was almost the same as the BiNoC_4VC(32) architecture, the added inter-router long wiring is always not cheap as technologies keep scaling down [27, 28]. Extensive researches considering layout-aware analysis of NoC reported that the overhead of NoC is huge due to not only the router area but also the inter-router wiring area [29–31]. Take area overhead for example, Pullini et al. [31] illustrated that, in 65 nm technology, a typical NoC having two 32-bit unidirectional links may need a 30 um-wide space to surround each processing element to provide routing space and insert buffers for long inter-router wires. Thus, in our point of view, rather than increasing the physical bandwidth by adding additional inter-router unidirectional links to a T-NoC architecture, we suggested using a BiNoC architecture where all unidirectional links were replaced by dynamic reconfigurable bidirectional links, to increase bandwidth utilization in practice.

6.6 Remarks

In this chapter, we clearly described the bandwidth utilization problem of conventional NoC design and proposed a novel concept of using bidirectional channel which can be adjusted based on real-time traffic requirement. A novel router architecture named BiNoC was provided with both wormhole and virtual-channel flow-control based mechanism. Then, an inter-router transmission scheme was provided first to achieve bidirectional data transmission. To avoid deadlock and starvation, a priority-based design of FSM was introduced. In addition, data transmission constraints including packet-ordering and packet transmission interruption were also described. Finally, both synthetic traffic pattern and real application benchmarks of E3S were provided to characterize BiNoC architecture. Extensive simulation results were provided in terms of various performance metrics and implementation overheads. All of these results illustrated that our proposed BiNoC architecture achieved better performance with low implementation overhead because of its better bandwidth utilization capability.

References

1. S. Kumar, A. Jantsch, J. P. Soininen, M. Forsell, M. Millberg, J. Oberg, K. Tiensyrja, and A. Hemani, "A Network-on-Chip Architecture and Design Methodology," in *Proceedings of the International Symposium on Very Large Scale Integration*, pp. 105-112, April 2000
2. R. Marculescu, U. Y. Ogras, L. S. Peh, N. E. Jerger, and Y. Hoskote, "Outstanding Research Problems in NoC Design: System, Microarchitecture, and Circuit Perspectives," *IEEE Transactions on Computer-Aided Design of Integrated Circuits and Systems*, vol. 28, no. 1, pp. 3-21, January 2009
3. T. Bjerregaard and S. Mahadevan, "A Survey of Research and Practices of Network-on-Chip," ACM Computing Surveys, vol. 38, no. 1, pp. 1-51, March 2006
4. D. Bertozzi, A. Jalabert, M. Srinivasan, R. Tamhankar, S. Sterqiou, L. Benini, and G. DeMicheli, "NoC Synthesis Flow for Customized Domain Specific Multiprocessor Systems-on-chip," *IEEE Transactions on Parallel and Distributed Systems*, vol. 16, no. 2, pp. 113-129, February 2005
5. U. Orgas, J. Hu, and R. Marculescu, "Key Research Problems in NoC Design: A Holistic Perspective," in *Proceedings of the International Conference on Hardware/Software Codesign and System Synthesis*, pp 69-74, September 2005
6. H. G. Lee, N. Chang, U. Y. Ogras, and R. Marculescu, "On-Chip Communication Architecture Exploration: A Quantitative Evaluation of Point-to-Point, Bus and Network-on-Chip Approaches," *ACM Transactions on Design Automation of Electronic Systems*, vol. 12, no. 3, pp. 1-20, August 2007
7. J. D. Owens, W. J. Dally, R. Ho, D. N. Jayasimha, S. W. Keckler, and L. S. Peh, "Research Challenges for On Chip Interconnection Networks," *IEEE Micro*, vol. 27, no. 5, pp. 96-108, November 2007
8. J. Lillis and C. Cheng, "Timing Optimization for Multisource Nets: Characterization and Optimal Repeater Insertion," *IEEE Transactions on Computer-Aided Design of Integrated Circuits and Systems*, vol. 18, no. 3, pp. 322-331, Mar. 1999
9. S. Bobba and I. N. Hajj, "High-Performance Bidirectional Repeaters," in *Proceedings of the Great Lakes Symposium on Very Large Scale Integration*, pp. 53-58, March 2000
10. A. Nalamalpu, S. Srinivasan, and W. P. Burleson, "Boosters for Driving Long Onchip Interconnects – Design Issues, Interconnect Synthesis, and Comparison with Repeaters," *IEEE Transactions on Computer-Aided Design of Integrated Circuits and Systems*, vol. 21, no. 1, pp. 50-62, January 2002
11. H. Ito, M. Kimura, K. Miyashita, T. Ishii, K. Okada, and K. Masu, "A Bidirectional- and Multi-Drop-Transmission-Line Interconnect for Multipoint-to-Multipoint On-Chip Communications," *IEEE Journal of Solid-State Circuits*, vol. 43, no. 4, pp. 1020-1029, April 2008
12. M. A. A. Faruque, T. Ebi, and J. Henkel, "Configurable Links for Runtime Adaptive On-Chip Communication," *in Proceedings of the Design Automation and Test in Europe Conference*, pp. 256-261, April 2009
13. M. H. Cho, M. Lis, K. S. Shim, M. Kinsy, T. Wen, and S. Devadas, "Oblivious Routing in On-Chip Bandwidth-Adaptive Networks," *in Proceedings of the Parallel Architectures and Compilation Techniques*, pp. 181-190, September 2009
14. Y. C. Lan, S. H. Lo, Y. C. Lin, Y. H. Hu, and S. J. Chen, "BiNoC: A Bidirectional NoC Architecture with Dynamic Self-Reconfigurable Channel," *in Proceedings of the International Symposium on Network-on-Chip*, pp. 266-275, May 2009

15. L. S. Peh and W. J. Dally, "A Delay Model and Speculative Architecture for Pipelined Routers," *in Proceedings of the International Symposium on High-Performance Computer Architecture*, pp. 255-266, January 2001
16. W. J. Dally and B. Towles, *Principles and Practices of Interconnection Networks*, Morgan Kaufmann, 2004
17. J. D. Owens, W. J. Dally, R. Ho, D. N. Jayasimha, S. W. Keckler, and L. S. Peh, "Research Challenges for On Chip Interconnection Networks," *IEEE Micro*, vol. 27, no. 5, pp. 96-108, Nov. 2007
18. D. A. Menasce and R. R. Muntz, "Locking and Deadlock Detection in Distributed Data Base," *IEEE Transactions on Software Engineering*, vol. SE-5, no. 3, pp. 195-202, May 1979
19. M. Lis, M. H. Cho, K. S. Shim, and S. Devadas, "Path-Diverse In-Order Routing," in *Proceedings of the International Conference on Green Circuits and Systems,* pp. 311-316, June 2010
20. S. Murali, D. Atienza, L. Benini, and G. DeMicheli, "A Multi-Path Routing Strategy with Guaranteed In-Order Packet Delivery and Fault-Tolerance for Network-on-Chip," in *Proceedings of the Design Automation Conference*, pp. 845-848, July 2006
21. M. Lis, K. S. Shim, M. H. Cho, and S. Devadas, "Guaranteed In-Order Packet Delivery using Exclusive Dynamic Virtual-channel Allocation," *Massachusetts Institute of Technology*, Technical Report, CSAIR-TR-2009-036, August 2009
22. D. C. Gazis, *Traffic Science*, John Wiley and Sons, 1974
23. R. Dick, "Embedded System Synthesis Benchmark Suites (E3S)," http://ziyang.eecs.umich.edu/~dickrp/e3s/, Accessed January 2011
24. M. Zhang and C. S. Choy, "Low-Cost VC Allocator Design for Virtual-channel Wormhole Routers in Network-on-Chip," in *Proceedings of the International Symposium on Networks-on-Chip*, pp. 207-208, April 2008
25. L. Shang, L. S. Peh, and N. K. Jha, "Powerherd: a Distributed Scheme for Dynamically Satisfying Peak-Power Constraints in Interconnection Networks," *IEEE Transactions on Computer-Aided Design of Integrated Circuits and Systems*, vol. 25, no. 1, pp. 92-110, January 2006
26. A. Banerjee, P. T. Wolkotte, R. D. Mullins, S. W. Moore, and G. J. M. Smit, "An Energy and Performance Exploration of Network-on-Chip Architectures," *IEEE Transactions on Very Large Scale Integration Systems*, vol. 17, no. 3, pp. 319-329, March 2009
27. R. Ho, K. W. Mai, and M. A. Horowitz, "The Future of Wires," *Proceedings of the IEEE*, vol. 89, no. 4, pp. 490-504, April 2001
28. *International Technology Roadmap for Semiconductors: Executive Summary*, Semiconductor Industry Association, 2007
29. F. Angiolini, P. Meloni, S. M. Carta, L. Raffo, and L. Benini, "A Layout-Aware Analysis of Networks-on-Chip and Traditional Interconnects for MPSoCs," *IEEE Transactions on Computer-Aided Design of Integrated Circuits and Systems*, vol. 26, no. 3, pp. 421-434, March 2007
30. I. Hatirnaz, S. Badel, N. Pazos, Y. Leblebici, S. Murali, D. Atienza, and G. DeMicheli, "Early Wire Characterization for Predictable Network-on-Chip Global Interconnects," *in Proceedings of the International Workshop on System Level Interconnect Prediction*, pp. 57-64, March 2007
31. A. Pullini, F. Angiolini, S. Murali, D. Atienza, G. DeMicheli, and L. Benini, "Bringing NoCs to 65 nm," *IEEE Micro*, vol. 27, no. 5, pp. 75-85, September 2007

Chapter 7
Quality-of-Service in BiNoc

A QoS-aware BiNoC architecture is proposed in this chapter to support guarantee-service (GS) traffic while reducing packet delivery latency. With the dynamically self-reconfigured bidirectional communication channels incorporated in BiNoC, as presented in Chap. 6, our proposed QoS-aware technique can promise more flexibility for various traffic flow patterns. Specifically, a novel inter-router communication protocol is proposed to prioritize bandwidth arbitration in favor of high-priority GS traffic flows. Multiple virtual-channels with prioritized routing policy are also implemented to facilitate data transmission with QoS considerations. Combining these architectural innovations, the QoS-aware BiNoC architecture can promise reduced latency of packet delivery and a more efficient channel resource utilization.

7.1 QoS Control in NoC

Most of today's consumer devices include a portion of real-time work flows with various QoS requirements that have to be considered in advance. Otherwise, violation of a specific real-time constraint may cause substantial performance degradation. Thus, providing different priorities to different applications, clients, or work flows to meet the multiple levels of guarantees become an urgent need in NoC design recently. In order to manage the on-chip network resources adequately, traffic flows are categorized into guaranteed service (GS) and best effort (BE) classes [1]. GS traffics are often used in timing critical signals or data streams such as the interrupt signal of the processors or a multimedia application. In contrast to GS traffics, BE traffics, which are generally applied in non-critical traffic flows, ensure transmission correctness and could only get a grant to use the bandwidth that the GS traffic did not need. Various algorithms facilitating the network performance under GS and BE traffics have been proposed for general computer networks [2]. However, they are not appropriate for on-chip

S.-J. Chen et al., *Reconfigurable Networks-on-Chip*,
DOI: 10.1007/978-1-4419-9341-0_7,

communication since their hardware implementation complexity and long computation period will become critical constraints.

As illustrated in previous section, the state-of-the-art QoS mechanisms for NoC can be categorized into two types: *connection-oriented* and *connection-less*. The quantitative modeling and comparison of these two schemes by running simulations on a multimedia platform have been presented in [3]. The results show that under a variable-bit-rate application, connection-less technique provided a better performance in term of the end-to-end packet delay. As pointed out in [4], connection-oriented network required resource reservation for the GS traffic in the worst-case and usually caused a huge amount of waste resources. Therefore, in this chapter, we propose a connection-less QoS mechanism for our BiNoC architecture to achieve better service and efficient dynamic resource allocation.

7.2 Typical Connection-Less QoS Mechanism for NoC

A typical Connection-less scheme aggregates traffics into different QoS requirements and allocates the traffics with different priorities such that they can be easily adapted to a virtual-channel flow-control based NoC router. Each virtual-channel can be seen as one individual buffer queue that was appropriately prioritized. Thus, only the traffic with a particular priority can be assigned to the corresponding virtual-channel at the VA stage.

Considering a typical connection-less QoS design in an NoC router, the SA module collects the channel request from each input virtual-channel, and processes the intra-router arbitration to resolve the request contention in the local router. While a contention involving GS and BE traffics occurs, the SA module will ignore the requests from the BE traffics and allocate the channel for the GS traffics. Figure 7.1 shows an example of intra-router arbitration at SA.

The first-stage arbitration of SA chooses at most one channel request from each input unit. Thus, GS traffic at VC0 has a higher priority to enter the second-stage arbitration phase and strive for the output channel towards their desired directions. The second-stage arbitration shows that output1 is contended by two GS traffics and output2 is contended by one GS traffic and one BE traffic. Since there are two GS traffics contending for output1, an arbitration technique such as round-robin is needed to be performed. On the other hand, the BE traffic from input3 has to wait for the GS traffic from input1. Then, the winner at the second-stage can get into the ST stage at the next cycle.

7.3 Motivational Example

Considering the conventional unidirectional NoC router architecture, all of the resource contentions between different traffics with various QoS requirements are arbitrated at a local router. Figure 7.2a illustrates the intra-router arbitration for

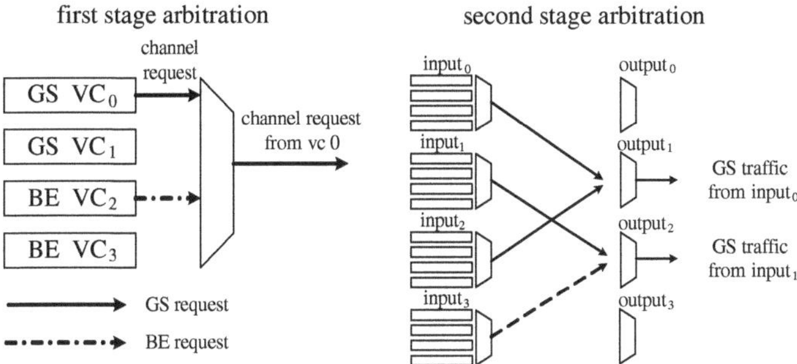

Fig. 7.1 Typical connection-less QoS design in an NoC router

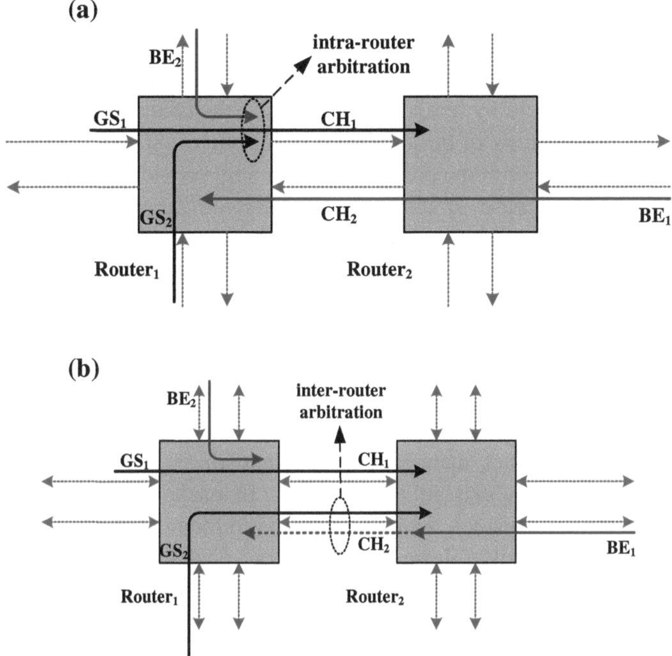

Fig. 7.2 Intra-router arbitration with QoS requirements

traffics with different QoS requirements in a conventional unidirectional NoC architecture, while Fig. 7.2b illustrates potential reward which can be gained by the inter-router arbitration.

As illustrated in Fig. 7.2a, where two GS traffic flows (GS_1 and GS_2) and one BE flow (BE_2) are contending for the outgoing channel at $Router_1$ which can only allow one GS traffic get the channel resource and transmit to $Router_2$. In this case, only GS_1 is granted while another unidirectional channel with an opposite direction is used by the BE_1 traffic with a lower QoS requirement from $Router_2$.

However, if the channel direction can be dynamically adjusted according to the real-time traffic QoS requirement, the traffic with a higher QoS requirement will have less chance to be blocked. In contrast to the intra-router arbitration, the inter-router arbitration decides which traffic has a higher priority to use the bidirectional channel between each pair of neighboring routers. Back to the previous example as illustrated in Fig. 7.2b, the inter-router arbitration can be applied to arbitrate the authority of CH_2 between GS_2 and BE_1. Finally, the GS_1 and GS_2 traffics with higher QoS requirements can be simultaneously transmitted, thus a higher channel utilization for the GS traffic can be obtained.

7.4 QoS Design for BiNoC Router

The BiNoC architecture, as presented in Chap. 6, can dynamically self-reconfigure its channel direction based on real-time traffic need. The basic concept is to utilize the waste idle bandwidth by temporarily reversing the channel direction in BiNoC to relieve the congestions in the opposing traffic direction. However, the BiNoC architecture as explained in the previous section only provides good latency results for the BE traffic because of its better channel utilization flexibility, but it is incapable to support critical communication guarantees that are much more important for real world applications.

7.4.1 Prioritized VC Management and Inter-Router Arbitration

A flexible virtual-channel management mechanism is applied to enhance the authority of the GS packets in this chapter. In each direction of the router, a four-entry prioritized virtual-channel module is implemented. Two of the virtual-channels are specifically designed for GS packets but the other two virtual-channels can be utilized by both GS and BE packets, thus we can reduce the blocking probability of GS packets in the VA stage. The intra-router arbitration is applied to the prioritized virtual-channels at the SA stage according to the QoS requirements of traffic flows.

To maximize the advantages of dynamically reconfigurable bidirectional channels in BiNoC and provide a superior performance guarantee for the GS packets, an inter-router arbitration scheme is proposed and integrated into the channel-direction control protocol to extend the scope of the channel arbitration from a local router to adjacent routers. In this scheme, the GS packets will be assigned a higher priority in adjusting the channel direction such that we can further improve the communication performance of GS traffic.

Figure 7.3 shows the proposed connection-less QoS scheme in our router model. In order to keep up with the double bandwidth obtainable in each output direction, the number of intra-router channel arbiters is also doubled to pick two

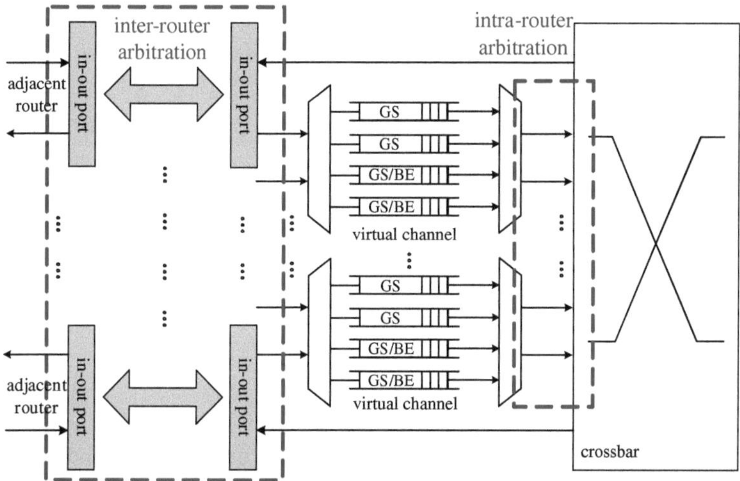

Fig. 7.3 Proposed QoS scheme in a BiNoC router

requests in each input direction. In this design, at most two GS packets requesting for the same output direction can be delivered simultaneously by configuring both bidirectional channels to the target direction without any blocking. Therefore, the contentions among GS packets from different queues can be greatly reduced.

In this QoS-aware BiNoC architecture, we give the GS packets a precedence over BE packets not only during the intra-router arbitration of other BE packets in the local router, but also in the inter-router arbitrater which tackles the channel direction with the adjacent routers. The detailed inter-router arbitration scheme for bidirectional channels will be illustrated in the next section.

7.4.2 Prioritized Deadlock-Free Routing Restriction

In order to leave more available communication bandwidth for the GS traffic than the BE traffic in a bidirectional channel network, we propose a prioritized routing restriction for GS and BE traffics. Specifically, we restrict the BE traffic with a deterministic routing policy but apply an adaptive routing algorithm for the GS traffic. Since two of the virtual-channels are shared among GS and BE packets, the usage of the BE packets will be limited and result in a greater authority for the GS packets to utilize the network resources. Under this prioritized routing restriction, the GS traffic can adaptively adjust its routing path to avoid congested area in the network and prevent possible blocking effect caused by the BE traffic. Moreover, this adaptiveness of the GS packets can also reduce the switching frequency of a bidirectional channel by selecting the channel that is in a desirable direction for output.

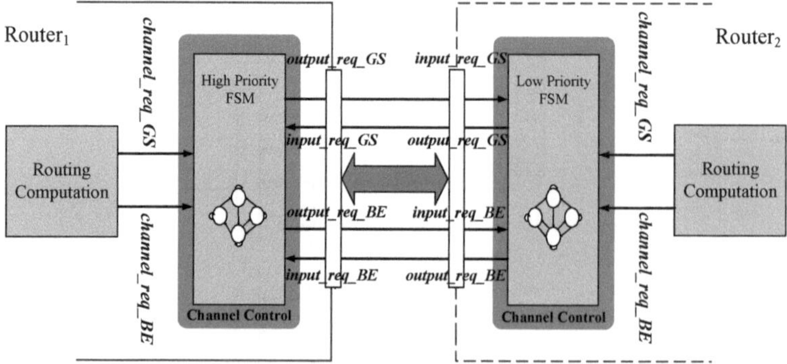

Fig. 7.4 Inter-router data transmission scheme

However, arbitrarily applying a deterministic routing algorithm and an adaptive routing algorithm to different traffic flows on the same network will cause additional deadlock problem. In order to conserve the deadlock-free property, we use an *Odd–Even* routing algorithm [5] and an *OE-fixed* routing algorithm [6], respectively, for the GS traffic and the BE traffic. *OE-fixed* is indeed a deterministic version of *Odd–Even* by removing the *Odd–Even*'s adaptiveness.

7.5 Inter-Router Transmission Scheme

The data transmission direction of a bidirectional channel needs to be dynamically configured according to the real-time QoS requirements of various traffics. To achieve this goal, the original inter-router data transmission scheme for one of the two bidirectional channels between each pair of adjacent routers, which has been presented in Chap. 6, should be modified as illustrated in Fig. 7.4.

The *Channel Control* module decodes the request and allocates the channel resource according to the QoS requirement of a traffic flow. Configuration of a bidirectional channel direction is controlled by a pair of FSMs in the *Channel Control* modules of the routers at both ends. The two FSMs exchange control states through a pair of signals: input request (*input_req_GS, input_req_BE*) and output request (*output_req_GS, output_req_BE*). One output request signal is pulled up when the sending-end router has data packet to transmit. Signal *output_req_GS* = 1 represents that the transmitting data is a GS packet; similarly, *output_req_BE* = 1 represents that the transmitting data is a BE packet. The output request signals from one router become the input request signals to the FSM of the other router. Each FSM also receives a channel-request (*channel_req_GS* or *channel_req_BE*) signal from the internal RC module. The channel-request signal is enabled when a data packet in the local router requests the current channel for forwarding data. Signal *channel_req_GS* = 1 represents that the local requestor is a GS packet, and *channel_req_BE* = 1 represents that the requestor is a BE

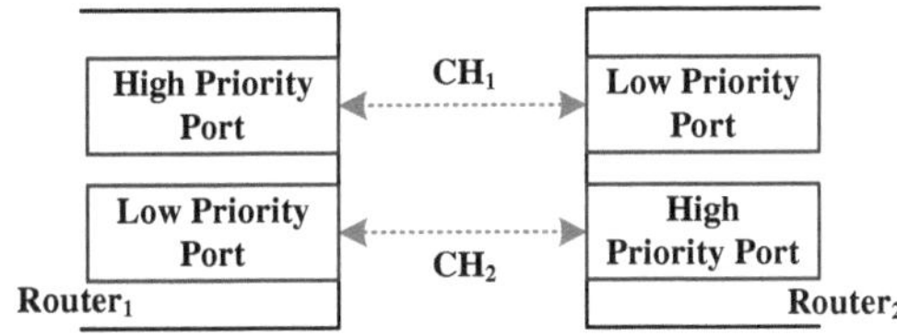

Fig. 7.5 Fixed channel prioritization mechanism

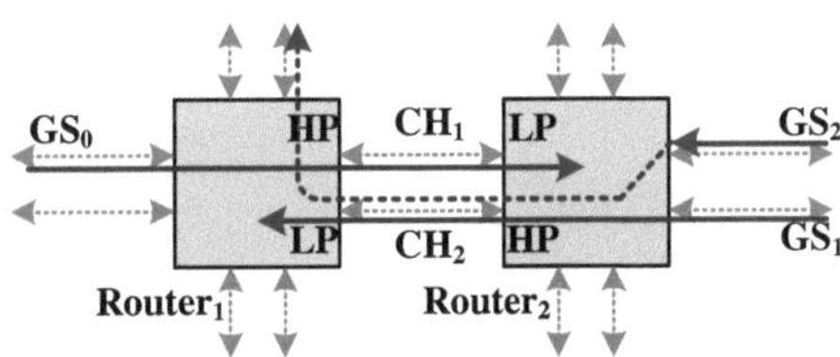

Fig. 7.6 Demonstration of starvation avoidance

packet. However, if the downstream input buffer is full, the channel request will be disabled. Each FSM also has an internal counter to keep track of the propagation delay through the synchronous channel. Note that all the input/output ports in a router are registered to provide clean signals from router to router in our design, as an NoC hardware implementation may require long wires. Therefore, two additional clock cycles are required to send a data packet out of the router.

Applying bidirectional channels in an NoC may cause an additional inter-router starvation problem, while the bidirectional channels are under contentions of traffic flows with the same QoS requirement and always being used by neighboring routers in opposite directions. To prevent the inter-router starvation problem, one of the two FSMs will be designated with a High-Priority (HP) and the other with a Low-Priority (LP). To enforce fairness of data transmission, the FSMs on the two ends of a channel (and between the same pair of routers) will be designated with opposite priorities as illustrated in Fig. 7.5. Under this constitution, if two traffic flows with the same QoS requirement want to transmit in opposite directions using the same bidirectional channel, the traffic flow at the high-priority terminal of the channel will be granted first.

Figure 7.6 illustrates the example where three traffics (GS_0, GS_1, and GS_2) with the same QoS requirement are contending for CH_1 and CH_2 simultaneously. We can find that GS_0 will be transmitted via CH_1 but GS_1 and GS_2 will be transmitted via CH_2, thus no inter-router starvation will happen.

7.6 QoS Design for BiNoC Channel-Direction Control

The state transition diagrams of a QoS High-Priority (HP) and a Low-Priority (LP) channel-direction control FSMs are shown in Figs. 7.7 and 7.8, respectively. Note that each FSM consists of four states: *free, idle, GS_wait,* and *BE_wait,* which are defined as follows:

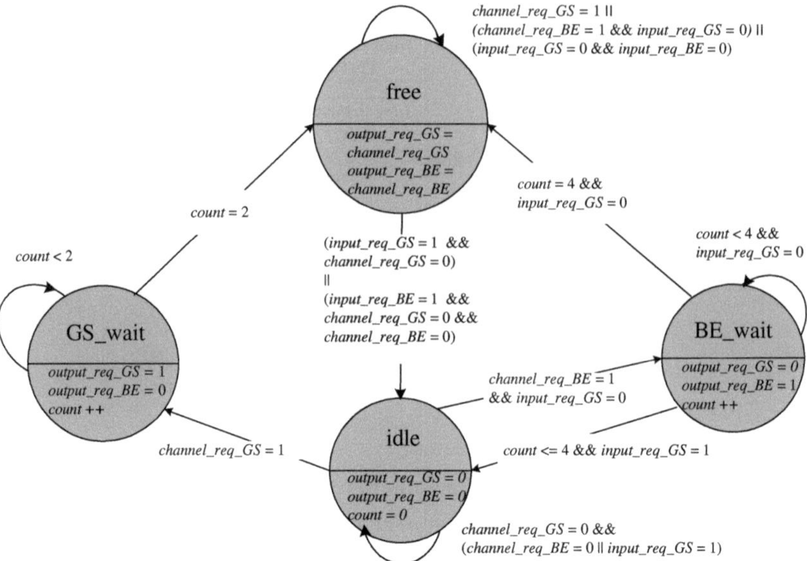

Fig. 7.7 QoS FSM for HP port of bidirectional channels

1. *free state*: The channel is available for data *output* to the adjacent router.
2. *idle state*: The channel is ready to *input* data from the adjacent router.
3. *GS_wait state*: An intermediate state for GS packets preparing the transition from the *idle* state with an input channel direction to the *free* state with an output channel direction.
4. *BE_wait state*: An intermediate state for BE packets preparing the transition from the *idle* state with an input channel direction to the *free* state with an output channel direction.

The operations of the HP FSM and the LP FSM are discussed in the following subsections.

7.6.1 High-Priority FSM Operations

As shown in Fig. 7.7, the High-Priority (HP) FSM will be initiated at a *free* state. It will remain at this state if no channel preemption is invoked by a higher priority packet from the neighbor router. In other words, as long as there is a data packet with a certain priority in the current router to be sent via this channel, the channel direction will not be influenced by the neighbor router unless the neighbor requestor has a higher priority. Even there are no data to transmit, if there is no request to send data from the other router, the channel direction shall still remain unchanged. The only condition that the HP FSM will leave the *free* state and enter

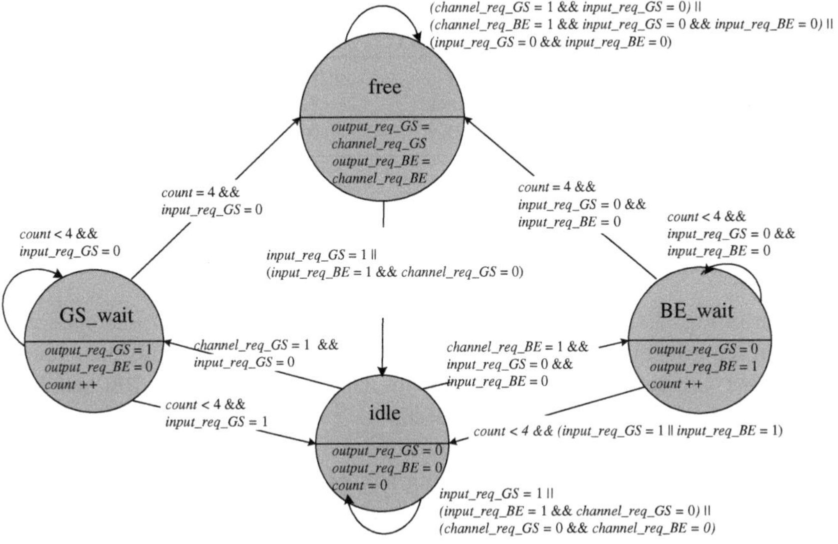

Fig. 7.8 QoS FSM for LP port of bidirectional channels

an *idle* state is when the channel authority is seized by a higher priority requestor from the neighbor router, or there is data to be sent from the neighbor router while none from the local end. At the *free* state, the output request signals are changed by the value of the channel requests. Hence the output value may become 0, allowing the other router to request the channel.

Once the FSM enters an *idle* state, it will remain at that state as long as there is no data to be transmitted outbound, or the incoming request from the neighbor router has a higher priority than the local channel request does. Moreover, while in the *idle* state, the output request is disabled. As soon as a GS channel request from any of the input units is detected (*channel_req_GS* = 1), the channel state will be triggered to a *GS_wait* state. The channel state will change to a *BE_wait* if a BE channel request is detected (*channel_req_BE* = 1) and no GS input request is received (*input_req_GS* = 0).

At each of the *wait* state, the output request is pulled up and a cleared counter increased. For the GS packets in the port with an HP FSM, it will definitely retrieve the channel authority after a two-cycle waiting process at the *GS_wait* state. As soon as the counter reaches two (*count* = 2), the HP FSM returns to the *free* state and starts GS data transmission. The purpose of the *GS_wait* state is to allow the GS output request signal (*output_req_GS* = 1) to reach the neighbor router such that the LP FSM will yield the channel by entering its *idle* state. At the *BE_wait* state in an HP FSM, the waiting period is four clock cycles considering the higher prioritized requestor from the neighbor router. The waiting process in the *BE_wait* state can only be interrupted by the GS input request from the neighbor router (*input_req_GS* = 1), while the BE input request will be neglected.

Table 7.1 NoC architectures used in our experiments

Architecture	NoC_QoS	BiNoC_4VC	BiNoC_QoS	BiNoC_QoS_OE
QoS mechanism	GS/BE	None	GS/BE	GS/BE
Total	5-in 5-out	10-inout	10-inout	10-inout
Channels routing	XY/XY	XY/XY	XY/XY	OE/OE-Fixed

7.6.2 *Low-Priority FSM Operations*

As shown in Fig. 7.8, the Low-Priority (LP) FSM is initiated at an *idle* state with the output requests disabled. If the HP FSM of the other router yields the channel, the LP FSM will leave the *idle* state for the *GS_wait* state or the *BE_wait* state according to the traffic type of the requestor.

Being a lower priority-end of data transmission, the LP FSM will undergo the waiting states for 4 clock cycles. If during any of these four cycles, the HP FSM requests the channel with the same or a higher priority compared to the local requestor, the LP FSM will return to the *idle* state. However, the GS requestor in the LP FSM still has the precedence over the BE requestor from the neighbor HP FSM. In other words, any of the input requests received during the *BE_wait* state will interrupt the waiting process of the BE requestor, while only the GS input request from the neighbor HP FSM can preempt the channel bandwidth and disrupt the local *GS_wait* state to an *idle* state. After experiencing four cycles ($count = 4$) without any interruption, the LP FSM will enter the *free* state and start transmission.

The state transition behavior of the *free* state depends on the QoS requirement of the traversing packet. According to the priority rule of the inter-router arbitration in our design, the transmission process of the local router will be halted by a higher prioritized request from the neighbor node. However, different from the HP FSM, an LP FSM may lose the output authority even if the priority level of the incoming request equals the outbound request transmitted by the local node.

7.7 Performance Evaluation

The detailed simulation environments are described in Appendix A. In this section, comprehensive comparisons will be provided to prove that our proposed QoS-aware BiNoC architecture can efficiently adjust the channel direction thus improve the performance of GS traffics based on the real-time traffic requirements. Table 7.1 lists the four router architectures used in our experiments. NoC_QoS represents a conventional QoS design in unidirectional NoC with four prioritized virtual-channels in each router direction. For each direction, one input channel and one output channel were used to connect neighboring routers. On the other hand, three BiNoC platforms were implemented with bidirectional channels.

The BiNoC_4VC, as presented in Chap. 6, using four virtual-channels in each direction and without any QoS mechanism, was implemented for comparison.

In our proposed BiNoC_QoS and BiNoC_QoS_OE, both the intra-router and the inter-router QoS mechanisms were applied to strengthen the communication performance of the GS traffics. Note that BiNoC_QoS_OE also adopted the proposed prioritized deadlock-free routing which routed the GS packets and BE packets respectively with adaptive *Odd–Even* and deterministic *OE-fixed* algorithms. All of the NoC architectures above were implemented in a four-stage pipelined router design based on virtual-channel flow-control, with four virtual-channels in length of 8 flits in each direction.

Three types of traffic patterns were run: uniform, transpose, and hotspot. The detailed behaviors of these three types of traffic are described in Appendix A. In the experiments, each traffic type of packets was generated with a varying injection rate for 25,000 cycles.

7.7.1 Comparison Between BiNoC_QoS and BiNoC_4VC

Figure 7.9 illustrates comparisons of BiNoC_4VC Architecture with BiNoC_QoS architecture in terms of latency versus flit injection rate. Results were obtained by running 7.9a uniform, 7.9b transpose, and 7.9c hotspot traffics, in which GS traffic occupied 20% of the total traffic.

The latency results were also compared between a BiNoC_4VC which only provided BE service, and a BiNoC_QoS which permited the GS packets having a higher priority to utilize the network resources. Since the prioritized intra-router and inter-router arbitration schemes favor the GS traffic on resource allocation, there was a significant amount of latency reduction in the GS packets while the BE packets got a worse performance. We can observe that the total average latency of BiNoC_4VC lay between the latency results of the prioritized GS and BE traffics. Note that the available virtual-channels and buffer space for BE packets may be occupied by GS packets under the applied QoS mechanism. Therefore, the network capacity for the BE packets is reduced such that the packet injection saturation point lessens. While the GS packets not only retained the authority for all the network resources but also had the precedence to use the channel bandwidth, the latency curves of GS packets were not increased drastically as the BE packets did, thus lengthening the packet injection saturation points.

The load-balanced nature of the uniform traffic caused a slight amount of GS packets not to conflict with each other frequently, thus resulting in a desirable average latency that did not explode to an extremely high value as shown in Fig. 7.9a. Since the transpose traffic is inherently suitable for BiNoC architectures with XY routing as presented in Chap. 6, the high-priority GS packets can preemptively utilize double bandwidth, thus the latency curve of the GS packets under transpose traffic in Fig. 7.9b was greatly improved. While in the hotspot traffic as illustrated in Fig. 7.9c, the performance enhancement of the GS packets in

(a)

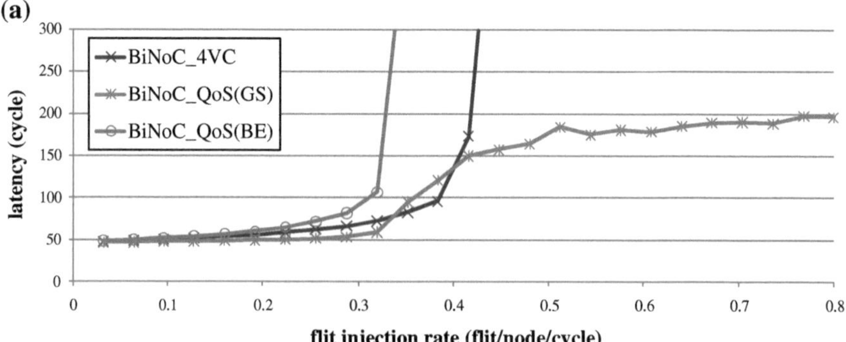

(b)

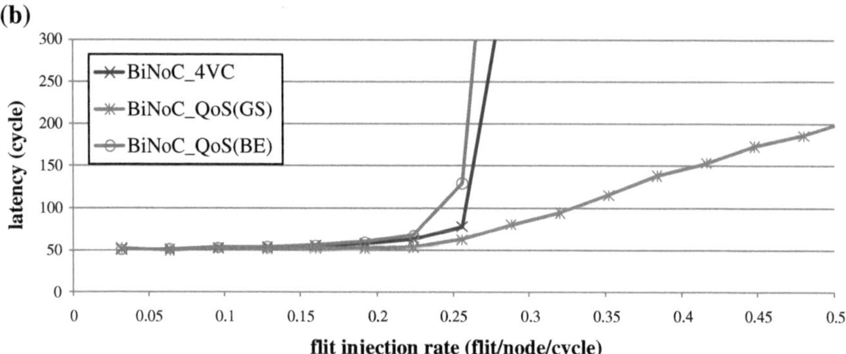

(c)

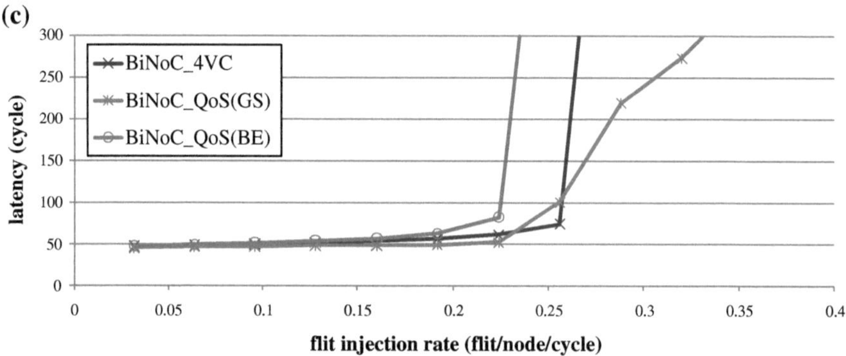

Fig. 7.9 Comparison of BiNoC_4VC with BiNoC_QoS

BiNoC_QoS architecture was still better than the BiNoC_4VC that cannot provide GS service, but limited by the congestion incurred on some specific nodes with a high channel load.

7.7.2 Comparison Between BiNoC_QoS and NoC_QoS

Figure 7.10 illustrates the comparison of NoC_QoS architecture with BiNoC_QoS in terms of latency versus injection rate. Results were obtained by running 7.10a uniform, 7.10b transpose, and 7.10c hotspot traffics, in which GS traffic occupied 20% of the total traffic. We can find that the benefit of our proposed QoS mechanism was that it can adjust channel direction dynamically according to the real-time QoS requirements of traffic flows. Since BiNoC_QoS with bidirectional channels had a better channel bandwidth utilization flexibility than the typical NoC_QoS, the communication performances of both GS and BE traffics under BiNoC_QoS architecture were improved simultaneously. Moreover, because of the extra prioritized inter-router arbitration, the latency results of the BE packets in BiNoC_QoS under transpose traffic and hotspot traffic even outperformed the GS packets in NoC_QoS as shown Figs. 7.10b, c. This is because the bidirectional channels can greatly improve the communication efficiency especially in the load-imbalanced traffics. In uniform traffic, the performance improvement was not obvious as shown in Fig. 6.10a since uniform traffic sends packets randomly and causes the bidirectional channels switch frequently. However, the switching of channel direction takes several cycles spending on the waiting state where no data can be transmitted during this period.

Under the proposed channel-direction control protocol in BiNoC_QoS, the GS packets can preemptively disrupt the BE packets and switch the channel direction for their transmission by the inter-router arbitration; thus, the communication efficiency of the GS packets was further improved. Comparing to NoC_QoS, the latency promotion was apparent under transpose traffic and hotspot traffic where the performance gap between GS traffic and BE traffic was enlarged in BiNoC_QoS as shown in Figs. 7.10b, c.

7.7.3 Analysis of Prioritized Routing

To illustrate the benefit of our prioritized routing restriction for traffic flows with different QoS requirements, Fig. 7.11 shows the comparison of a BiNoC_QoS architecture with a BiNoC_QoS_OE architecture in terms of latency versus injection rate. Results were obtained by running 7.11a uniform, 7.11b transpose, and 7.11c hotspot traffics in which GS traffic occupied 20% of the total traffic.

As shown in Fig. 7.11a, XY routing performed better than the proposed prioritized routing under uniform traffic condition. This result was consistent with other results as reported in [5, 7], since the XY routing strategy maintains long-term evenness of uniform traffic while the adaptive routing algorithm only pursues short-term benefits. On the other hand, the transpose traffic operates a much more regular pattern where the network load is imbalanced under XY routing policies. Therefore, the latency results of BiNoC_QoS_OE with adaptive routing for GS

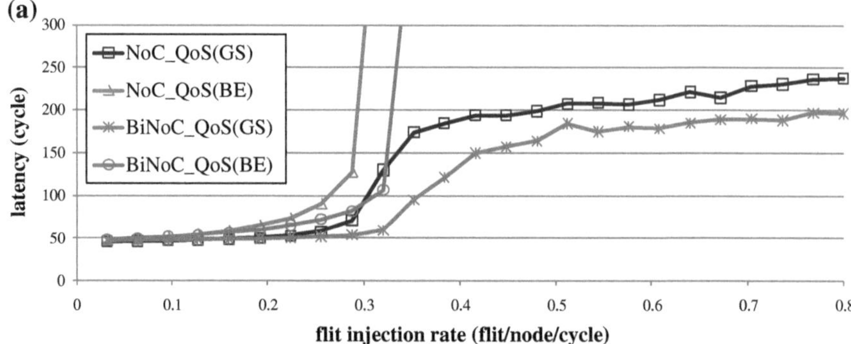

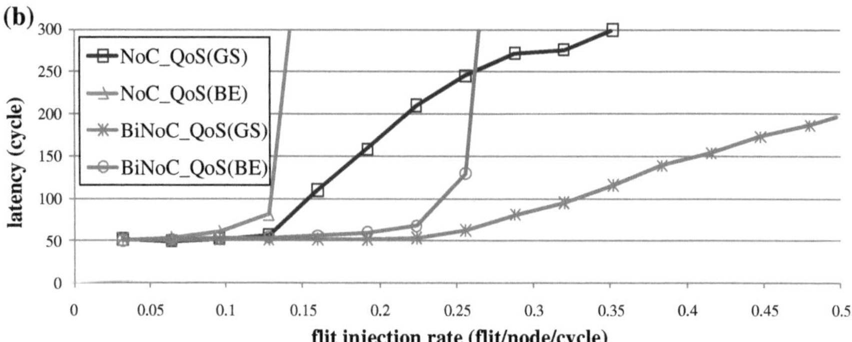

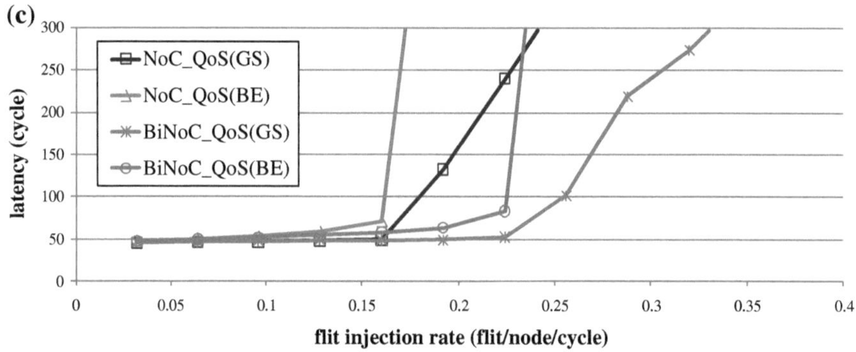

Fig. 7.10 Comparison of NoC_QoS with BiNoC_QoS

packets outperformed BiNoC_QoS significantly as illustrated in Fig. 7.11b. Neither uniform nor transpose traffic can simulate realistic application traffic conditions where multiple hotspots may be likely to occur. Figure 7.11c illustrates the latency results under hotspot traffic to demonstrate the consistent performance improvement.

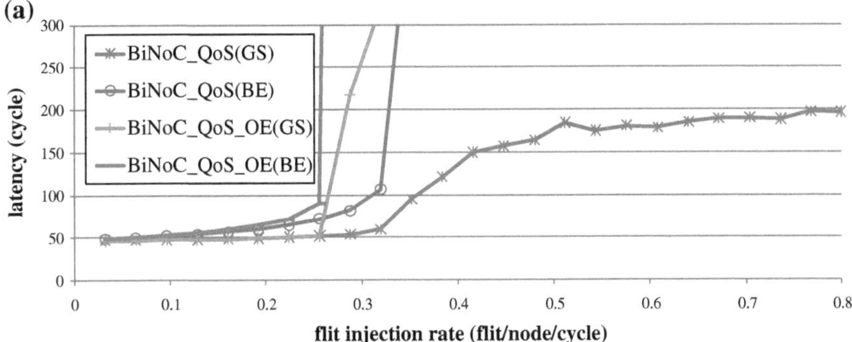

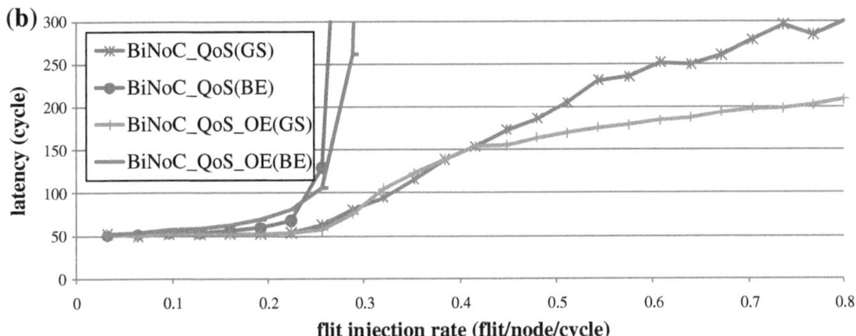

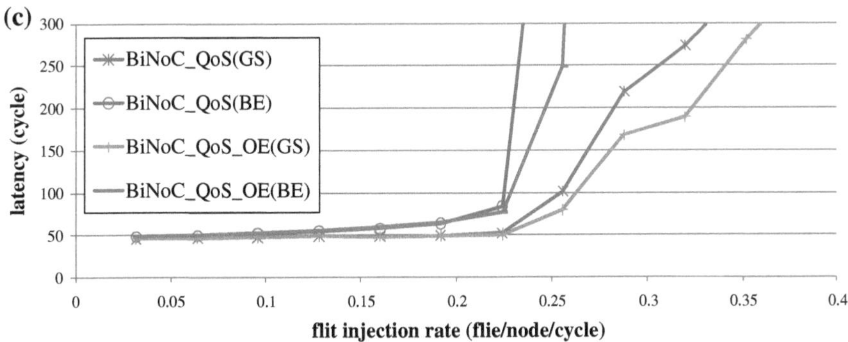

Fig. 7.11 Comparison of BiNoC_QoS with BiNoC_QoS_OE

Since the virtual-channel resources of the BE packets are shared with the GS packets, the BE traffics can also benefit by the performance improvement in the GS packets under the prioritized routing policy. Furthermore, the adaptive routing for the GS packets in BiNoC_QoS_OE can improve the latency results by dynamically selecting a routing path to evade several congested hotspots such that the data packets can be directly transmitted by the bidirectional channels that are in a *free* state.

(a)

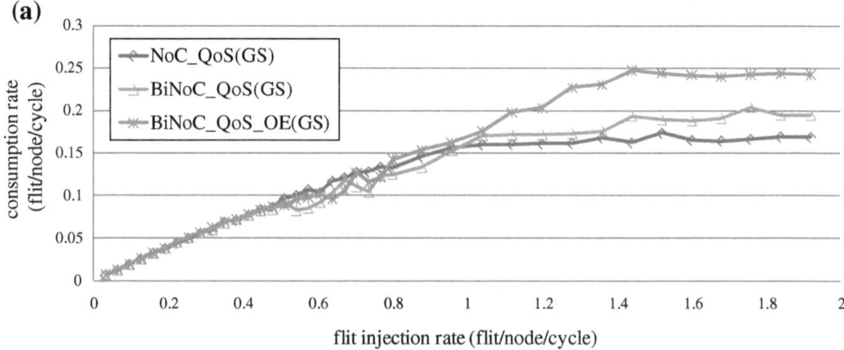

(b)

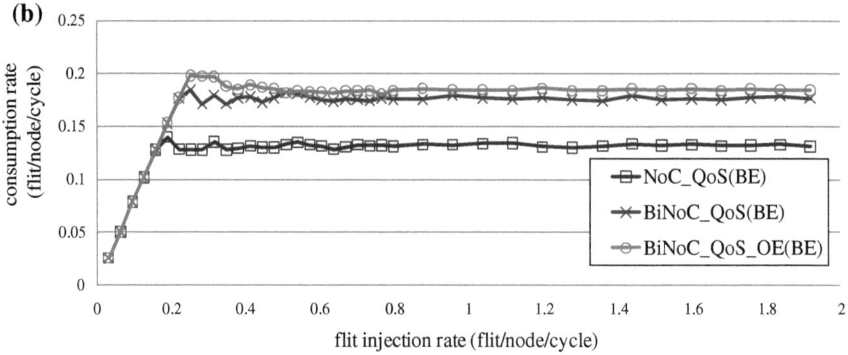

Fig. 7.12 Flit consumption rate analysis

7.7.4 Analysis of Consumption Rate

Figures 7.12a, b show the performance results of consumption rate of GS and BE traffics obtained by running hotspot traffic pattern with 20% GS ratio to the total traffic.

At low injection rates, the injection rates matched with the consumption rates of an NoC. But after reaching a definite injection rate, the flits were no longer being consumed at the same rate as they were being input to the NoC. This phenomenon of NoC injection rate increasing faster than their consumption rate is analogous to fundamental traffic model reaching critical density point [8]. Since the injection rate of GS traffic is one-fourth of the BE traffic, the slopes of the GS curves at low injection rates were smaller than the BE curves.

Note that the consumption rate of NoC_QoS was always worse than others due to the flexibility limitation of the unidirectional channels. As a result of the adaptive routing policy applied on the GS packets that can select a more suitable route to their destinations with fewer conflictions, BiNoC_QoS_OE achieved a significant improvement on flit consumption rate compared to BiNoC_QoS.

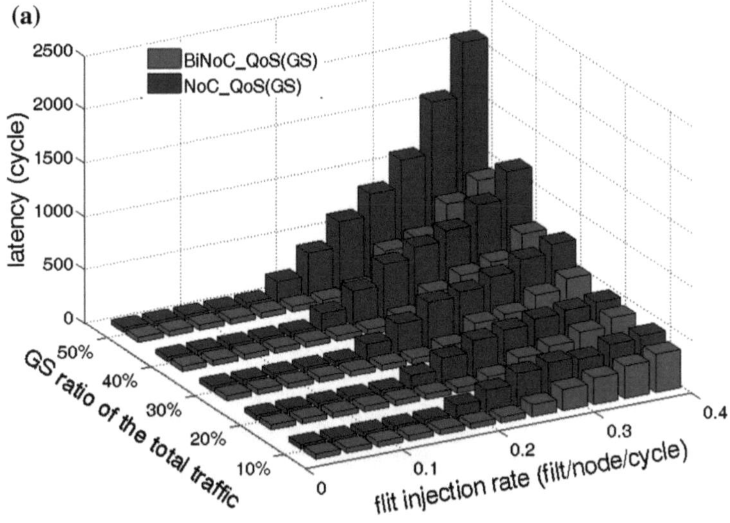

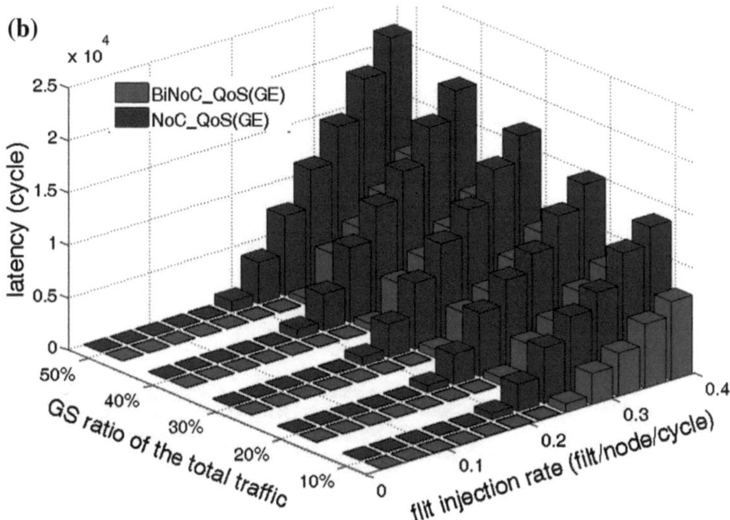

Fig. 7.13 Comparisons between NoC_QoS and BiNoC_QoS

7.7.5 Comparison Between GS and BE Traffics

Figures 7.13a, b illustrate the latency results of GS packets and BE packets respectively between NoC_QoS and BiNoC_QoS under various ratios of GS traffic to the total traffic. Results were obtained by running hotspot traffic. The simulation condition applied was hotspot traffic because it is more realistic compared to the

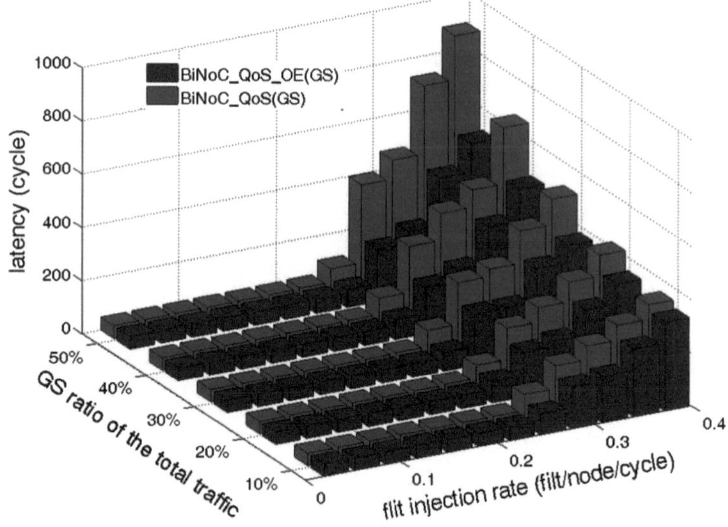

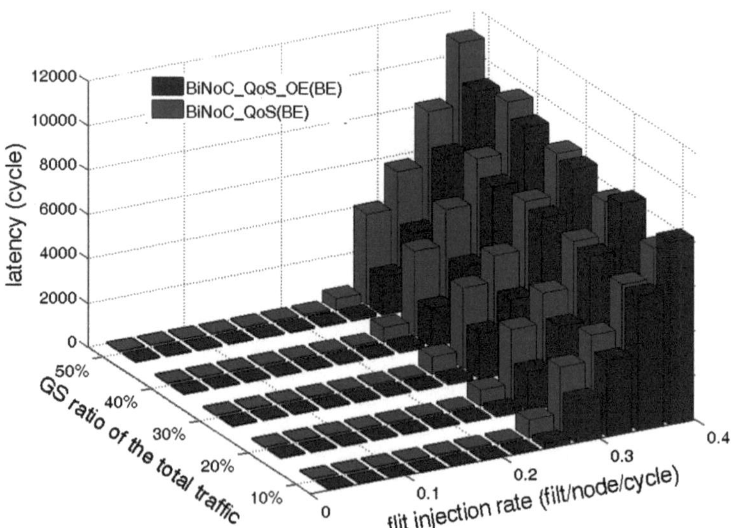

Fig. 7.14 Comparisons between BiNoC_QoS and BiNoC_QoS_OE

other two traffics. The GS ratio was varied from 10 to 50% of the total traffic in our experiments, because only a small portion of the total network traffic was expected to be QoS guaranteed in realistic cases.

Under a fixed flit injection rate, we can observe that as the ratio of GS traffic increased, the latency of GS packets stepped up as a result of the confliction between the GS traffics. However, the GS latency of BiNoC_QoS was not increased drastically as NoC_QoS did especially when the ratio grew higher than 30%. This is attributed to the flexibility of the bidirectional channels and the proposed inter-router arbitration scheme that greatly enhance the communication performance of the GS packets. As to the latency results of BE traffic, they also got worse as the GS ratio grew, since more network resource utilization priority was given to the GS traffic. We also plot the latency comparisons between BiNoC_QoS and BiNoC_QoS_OE for GS and BE traffic flows as shown in Figs. 7.14a, b. Results were obtained by running hotspot traffic. The GS packets with adaptive routing strategy applied in BiNoC_QoS_OE can better accommodate the hotspot traffic; thus, the GS traffic in BiNoC_QoS_OE always perform better than BiNoC_QoS under various injection rates and GS ratios. Since the GS packets in BiNoC_QoS_OE can dynamically select their routes to avoid congestion areas and occupied channels, they have a smaller possibility to be blocked at the network. While the GS packets spend a shorter latency to traverse the network, the duration that they occupy the network resources is also reduced, thus allow the BE packets have more chance to utilize the channel bandwidth at early periods. Therefore, the BE traffic in BiNoC_QoS_OE that applied deterministic OE-fixed routing still outperformed the BE traffic in BiNoC_QoS that used deterministic XY routing.

7.8 Remarks

In this chapter, a connection-less QoS mechanism based on the BiNoC backbone was presented. An inter-router channel-direction control protocol which assigns a higher priority for the critical GS traffic to traverse the network was proposed. Moreover, a flexible virtual-channel management mechanism and a novel prioritized routing policy were integrated in our QoS design to further enhance the communication efficiency of the GS packets. Extensive experimental results demonstrated that the proposed inter-router arbitration scheme can significantly increase the channel utilization for GS packets to get more promising performance guarantees than in a conventional NoC with intra-router arbitration. Additionally, the experiment results also proved that our prioritized routing policy can promote the overall performance by increasing the routing flexibility of the GS traffic to avoid possible traffic congestion.

References

1. W. J. Dally and B. Towles, Principles and Practices of Interconnection Networks, Morgan Kaufmann, 2004
2. R. Guerin and V. Peris, "Quality-of-Service in Packet Networks: Basic Mechanisms and Directions," in Proceedings of the Computer Networks, pp. 169-179, February 1999
3. M. D. Harmanci, N. P. Escudero, Y. Leblebici, and P. Ienne, "Quantitative Modeling and Comparison of Communication Schemes to Guarantee Quality-of-Service in Networks-on-

Chip," in Proceedings of the International Symposium on Circuits and Systems, pp. 1782-1785, May 2005
4. E. Rijpkema, K. G. W. Goossens, A. Radulescu, J. Dielissen, J. V. Meerbergen, P. Wielage, and E. Waterlander, "Trade-offs in the Design of a Router with Both Guaranteed and Best-Effort Services for Networks-on-Chip," in Proceedings of the Design Automation and Test in Europe Conference, pp. 350-355, March 2003
5. G. M. Chiu, "The Odd-Even Turn Model for Adaptive Routing," IEEE Transactions on Parallel and Distributed Systems, vol. 11, no. 7, pp. 729-738, July 2000
6. J. Hu and R. Marculescu, "DyAD–Smart Routing for Networks-on-Chip," in Proceedings of the Design Automation Conference, pp. 260-263, June 2004
7. M. Li, Q.A. Zeng, and W. B. Jone, "DyXY - a Proximity Congestion-Aware Deadlock-Free Dynamic Routing Method for Network on Chip," in Proceedings of the Design Automation Conference, pp.849-852, July 2006
8. D. C. Gazis, Traffic Science, John Wiley and Sons, 1974

Chapter 8
Fault Tolerance in BiNoC

For fault-tolerant data-link connections, a novel Bi-directional Fault-Tolerant NoC (BFT-NoC) scheme that supports both static and dynamic channel failures is proposed in this chapter. Except for a little performance loss, BFT-NoC can keep the system in normal operation when multiple communication channels are either permanently broken or temporarily failed in an on-chip network. In contrast to the conventional fault-tolerant schemes based on detouring packets, the operation of BFT-NoC is transparent to the adopted routing algorithm. That is, BFT-NoC can be more seamless and efficient since changing routing rules between normal and fault-tolerant operation modes is needless. Accordingly, it is possible for BFT-NoC to perform better in both resource utilization and application feasibility compared with other detour-based schemes.

8.1 Problem and Motivation

In the NoC data-link layer, the communication reliability could be strongly impacted by an unreliable router-to-router communication channel. Therefore, many fault-tolerant schemes, such as [1–5], have been proposed, most of which were conducted on detouring packets to avoid going through the possibly faulty channels. In such a way, either more restrictions on path selections or off-line reconfigurations in routing tables are required to avoid deadlocks during detouring packets. Correspondingly, in a fault-tolerant operation mode (i.e., faults exist in the network), on one hand, schemes with limited path selections always lead to inferior routing adaptivity and lower reliability; on the other hand, schemes based on reconfiguring global routing tables need additional table costs and cannot tolerate faults dynamically. Since deadlock freedom is a crucial problem in NoC fault-tolerant routing algorithms, the detour-based scheme is continually open to discussion.

S.-J. Chen et al., *Reconfigurable Networks-on-Chip*,
DOI: 10.1007/978-1-4419-9341-0_8,

Recently, while bidirectional channels have been used to enhance the NoC communication performance [6, 7], there has been no study that applied bidirectional channels in the fault-tolerant NoC research. In this chapter, we propose a novel scheme, named Bi-directional Fault-Tolerant NoC (BFT-NoC), which utilizes the bidirectional channel to deal with the fault-tolerance issues in the NoC data-link layer.

8.2 Fault-Tolerance Basics

With continued shrinkage of semiconductor process feature sizes, on-chip operating frequency and transistor density will continue to grow while the supply voltages will continue to decrease. In [8], it is commented that "*Electrical noise due to crosstalk, electromagnetic interference, and radiation-induced charge injection will likely produce data errors, also called upsets. Thus, transmitting digital values on wires will be inherently unreliable and nondeterministic*". Therefore, NoCs need to be designed with specific fault-tolerant schemes to counterbalance the negative impact of the increasing failure rate of on-chip interconnects.

8.2.1 Fault Types in NoCs

Along with the continued scaling in process technologies, the on-chip communication is not guaranteed error-freedom in future NoCs. The main challenge is represented by the increased prominence of noise sources such as power supply noise, crosstalk noise, inter-symbol interference, electromagnetic interference, thermal noise, and noise induced by alpha particles [9]. Consequently, failures probably exist in a chip and can be static or dynamic. Referring to the definition in [9], *static failures* are present in the network when the system is powered on and *dynamic failures* appear at random during the operation of the system. These two types of failures are respectively caused by *permanent faults* and *transient faults*. Alternatively, *transient faults* generally occur in the field and it has been showed that 80% of system failures are associated with *transient faults* [9]. For these reasons, fault-tolerant schemes are required to provide reliable communication services for the next generation of NoCs.

8.2.2 Fault-Tolerance in NoCs

Along with the shrinking process technology, both *permanent fault* and *transient fault* could happen in on-chip communication channels [9]. Accordingly, some studies have been proposed to tackle this emerging issue for the next generation of NoCs. An on-line fault diagnosis using a router to send "ADDRESS" flits and receive "ANSWER" flits between all of its neighbors to check the connection status was proposed in [1]. Besides, in Ref. [5], a built-in and self-diagnosis mechanism is used to yield fault information of the network. Especially, the fault diagnosis can dynamically execute and discriminate between *permanent* and *transient faults*. For deadlock free fault-tolerant routing, an on-line and turn-model based reconfigurable routing algorithm was proposed in [2] to support one faulty router in a mesh network. Another turn-model based and highly resilient routing algorithm provided in [3] can tolerate multiple faulty channels distributed in the network; but an off-line routing table reconfiguration must be performed to restart communications. In contrast to the turn-model based schemes, a virtual channel based fault-tolerant routing algorithm was given in [4]. However, these additional virtual channels are reserved for fault-tolerance and do not add to the performance of the system [3]; besides, the extra virtual channel implementation overheads [10] cannot be avoided.

8.2.3 Bidirectional Channels in NoCs

In conventional NoCs, a router uses a pair of unidirectional channels to connect with a neighboring router. One of the channels is used for transmission (TX) and the other is reserved for reception (RX). Compared with the unidirectional channel, a bidirectional channel allows to be dynamically configured to transmit data in either direction. Recently, applying bidirectional channels in increasing NoC communication performance has been demonstrated in [6, 7]. However, using bidirectional channels to enhance NoC fault-tolerance remains as a matter to be further discussed. Thus, it would be of interest to learn more about the capability and flexibility as well as the advantages and limitations regarding a bi-directional fault-tolerant NoC which will be discussed in details in the following sections.

8.2.4 Problems of Existing Fault-Tolerant Schemes

In conventional NoCs, a router uses a TX-Channel and a RX-Channel to communicate with its neighboring router as shown in Fig. 8.1a. When the TX-Channel is faulty as illustrated in Fig. 8.1b, it will lead to a fatal error if transmitting packets from router R1 to router R2 in a non-fault-tolerant system. Even applying existing fault-tolerant schemes which can detour packets through the other possible routing paths; there are still two major problems to be taken into consideration as follows:

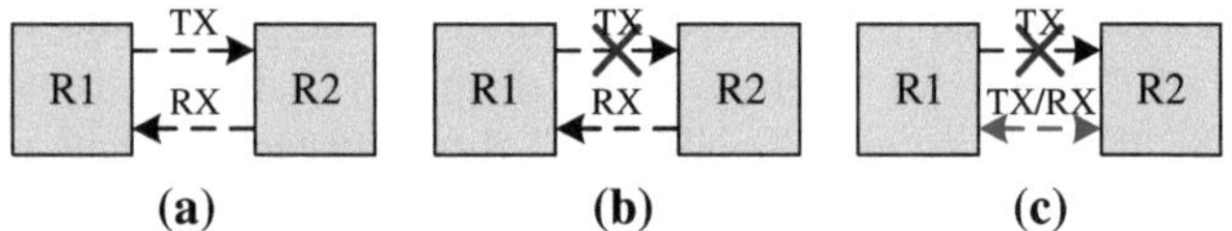

Fig. 8.1 Channel transmission directions and connection statuses of **a** non-faulty NoC, and a faulty channel in **b** conventional NoC and **c** the BFT-NoC

1. *Problem 1*: The alternative path does not exist (e.g., in an irregular topology network such as the SPIN in [8]).
2. *Problem 2*: The alternative path possibly leads to a deadlock (e.g., violating the turn rules of the adopted turn model [10]).

To the best of our knowledge, no NoC fault-tolerant scheme can handle *Problem* 1 before our proposal. For *Problem* 2, to avoid deadlocks in the fault-tolerant operation mode, schemes based on a modified existing routing algorithm such as [2] always have inferior routing adaptivity and lower reliability. Besides, some studies rely on re-configuring the global routing tables such that the system can restart after completing the re-configuration [3]. However, this kind of off-line schemes cannot dynamically handle faults as mentioned in the previous section.

8.2.5 Methodology of our Proposed Scheme

As shown in Fig. 8.1c, although the TX-Channel is faulty, the intact RX-Channel can be changed in bidirectional operations to provide both TX and RX communications. BFT-NoC can provide fault-tolerance except the case where we have a pair of faulty channels as shown in Fig. 8.2a. Therefore, we analyzed the probability that there are a certain number of faulty channels in an 8×8 mesh network with at least one pair of faulty channels located between two neighboring routers. Figure 8.2b shows that even when the number of faulty channels was increased to seven, the probability was less than 10%. In other words, our BFT-NoC provided above 90% reliability in an 8×8 mesh where there existed seven faulty channels. Furthermore, BFT-NoC can achieve almost 100% reliability after coupling with another detour-based scheme; we will discuss this reliability enhancement in detail in Sect. 8.3.7.

8.3 Proposed Bi-Directional Fault-Tolerant NoC Architecture

In this section, we propose a Bi-directional Fault-Tolerant NoC (BFT-NoC) scheme to control the two problems mentioned above by utilizing bidirectional channels.

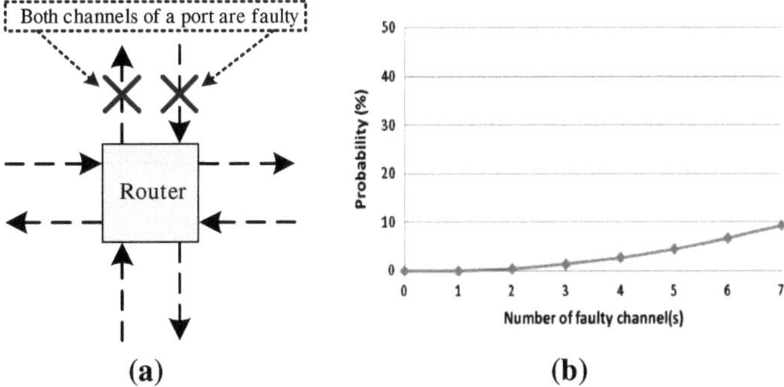

Fig. 8.2 **a** Pair of faulty channels example and **b** its probability in an 8 × 8 mesh

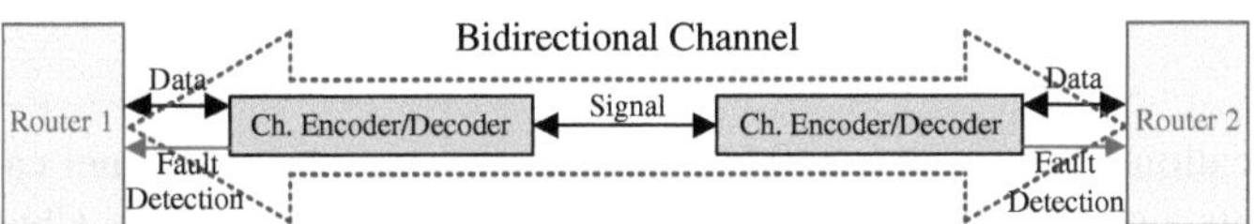

Fig. 8.3 Bidirectional channel example

8.3.1 Bidirectional Channels

In an NoC, faults in connections between routers could be permanent faults caused by electronic migration and dielectric breakdown or transient faults such as soft error and timing fault [9]. In these cases, a fault detection mechanism is required in an unreliable on-chip communication environment and the approach involves communications between a sender and a receiver. For example, Cyclic Redundancy Code (CRC) is a common channel error detection code as introduced in [9]. Since the CRC designs in both the channel encoder and the channel decoder are identical, it is feasible to implement a bidirectional channel with almost the same hardware cost of a unidirectional channel as shown in Fig. 8.3.

8.3.2 Bidirectional Router Architecture

Referring to Fig. 8.4, in a Bi-directional Fault-Tolerant NoC (BFT-NoC), two bidirectional channels are equipped between neighboring routers. An additional channel controller is required to dynamically select the data flow direction in the bidirectional channels whenever any faulty channel is detected or the faulty channel is recovered. In this chapter, we focus on studies about the capability and feasibility in applying the bidirectional channel for NoC fault-tolerant

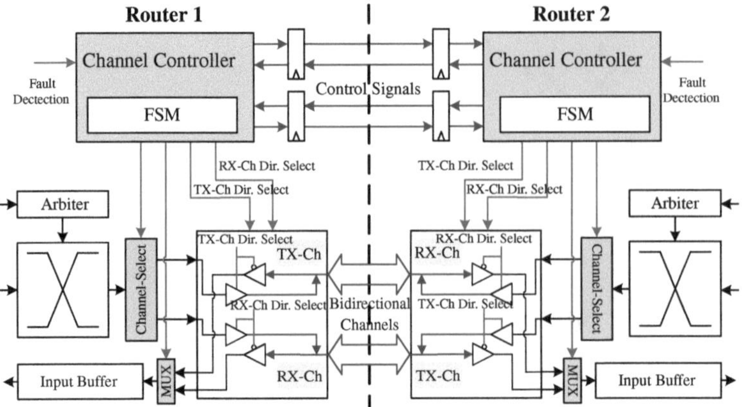

Fig. 8.4 Router architecture and inter-router connections

communications. Hence, in normal operations (i.e., no faulty channel case) in our experiments, one of the bidirectional channel pair is used for the TX-Channel (TX-Ch) and the other is reserved for the RX-Channel (RX-Ch). When one of the two channels is faulty, the other channel can be changed to operate bi-directionally; that is, the intact channel can be shared by both the TX and RX data flows. Compared with the other fault-tolerant schemes as mentioned above, our proposed mechanism can dynamically react to the faulty conditions and be independent of the adopted routing algorithm. Once both channels between neighboring routers are faulty, our designed channel controller reflects this condition and then can cooperate with other schemes such as [2, 3] to seek alternative routing paths. However, the failure rate in an on-chip environment is very small (as shown in Table 4.1 of [9]); accordingly, and referring to Fig. 8.2b, BFT-NoC can individually provide a high reliability when the number of faults is generally a small value. In the next section, we will introduce the details of our BFT-NoC mechanism.

8.3.3 Channel Direction Change Handshaking

To share a channel (ch) bandwidth between two neighboring routers, a handshaking protocol is required to avoid conflicts caused by transmitting data from both sides at the same time. From the viewpoint of a router, four control signals as shown in Fig. 8.5 are required and their definitions are given as:

1. *tx_req*: output, asserted when requesting to use the channel for TX.
2. *tx_gnt*: input, asserted when the TX request is granted.
3. *rx_req*: input, asserted when being requested to use the channel for RX.
4. *rx_gnt*: output, asserted when the RX request is granted.

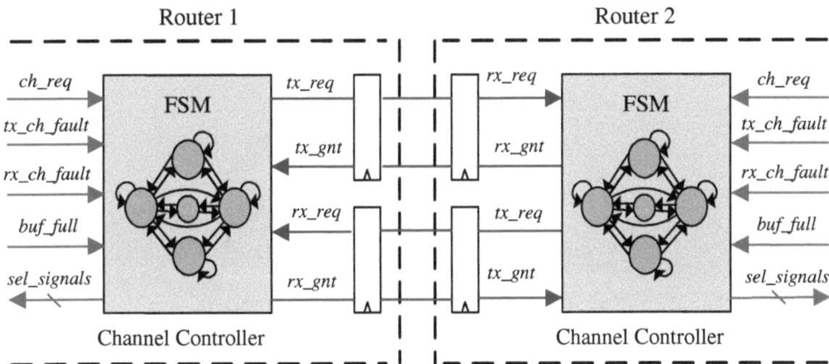

Fig. 8.5 Intra-router and inter-router control signals

It is noticeable that each of these four control signals has a corresponding definition in its counterpart router. Referring to Fig. 8.5, the *tx_req* in Router 1 corresponds to the *rx_req* in Router 2; and the *tx_gnt* in Router 1 corresponds to the *rx_gnt* in Router 2. Besides, some other control signals are used for the channel controller within a router. For example, *ch_req* is asserted when the bidirectional channel is requested to be used for transmitting packets. An asserted *tx_ch_fault* means that the TX-Channel is faulty, and the asserted *tx_ch_fault* could be de-asserted after the TX-Channel is recovered from a *transient fault*; similarly, *rx_ch_fault* indicates the fault status of the RX-Channel. The signal of *buf_full* is used to prevent *in-router deadlocks* which will be discussed later. Last, *sel_signals* are the selections of multiplexers and tri-state buffers as shown in Fig. 8.4.

8.3.4 Fault-Tolerance Control Procedure

The Finite State Machine (FSM) of the channel controller as depicted in Fig. 8.6 comprises five main states:

1. *Normal*: In the absence of faulty channel.
2. *TX*: The intact channel is available for transmitting data out.
3. *RX*: The intact channel is ready for receiving data in.
4. *Wait*: An intermediate state from the TX state to the RX state.
5. *Paired-Fault*: Both channels are faulty.

In Fig. 8.6, if a channel is detected faulty while the FSM of the channel controller is at the *Normal* state, the FSM will transfer to the *TX* or *RX* state depending on whether the faulty channel is a RX-Channel or a TX-Channel. When the FSM is at the *TX* state and receives a reception request (*rx_req*), then it can transfer to the *RX* state when the channel is idle or transfer to the *Wait* state to wait for the end of

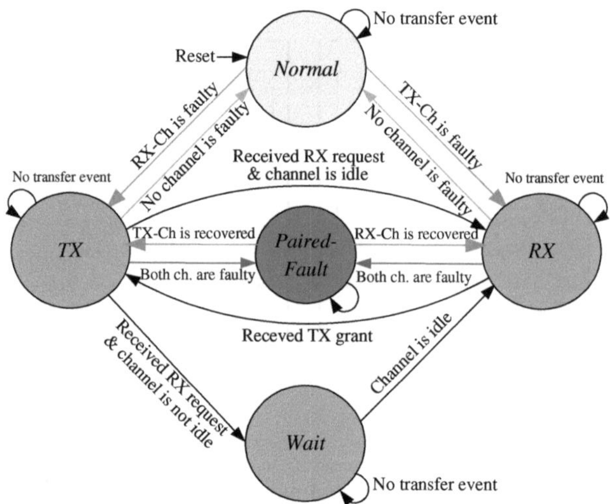

Fig. 8.6 Main states and state transitions in the final state machine

the current packet transmission. While transferring to the *RX* state, the reception grant (*rx_gnt*) is given to the counterpart channel controller. Otherwise, when the FSM is at the *RX* state, if the channel controller receives a channel request (*ch_req*) internally to demand outputting data, then it should assert the transmission request (*tx_req*) first, after receiving the transmission grant, *tx_gnt* (*i.e.*, the *rx_gnt* at the counterpart router); then the FSM can directly transfer to the *TX* state since the counterpart router has completed its data transmission. In case of detecting a pair of faulty channels, the FSM will transfer to the *Paired-Fault* state in order to inform a detour-based scheme such as in [2] and [3] to handle this faulty case. Especially, if the error is transient and whenever the faulty channel is recovered, the FSM will retrieve a *TX, RX or Normal* state.

8.3.5 In-Router Deadlock and its Solution

In our implementations, we adopted XY and Odd–Even [10] as the deadlock free routing algorithms. Although deadlock freedom can be guaranteed among routers in a mesh network, a new *in-router deadlock* phenomenon could exist in our proposed BFT-NoC with wormhole switching. As shown in Fig. 8.7a, there are two faulty channels respectively located in the west and east sides of the router, where both FSMs of the channel controllers are in the *RX* state and waiting for each other to transfer to the *TX* state to free up the buffer spaces for accommodating the remaining flits of the incoming packets. Unfortunately, the mutually waiting condition will continue forever and form a deadlock in the router. To relieve this *in-router deadlock* condition, we select to preempt the FSM from

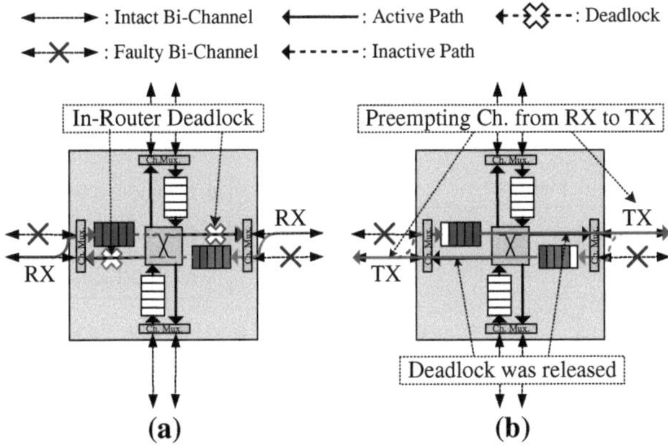

Fig. 8.7 **a** In-router deadlock condition and **b** its solution

the *RX* state to the *TX* state while the channel controller receives both a channel request (*ch_req*) and a buffer full condition (*buf_full*) as illustrated in Fig. 8.7b. Consequently, the buffer will be flushed and then the FSM can return to the *RX* state to continue receiving the residual flits.

8.3.6 Failure Rate Enhancement

In the traditional NoC with uni-directional data channel, the *channel failure rate*, FR_{Ch}, can be defined as the inverse of the channel's *mean time between failures*, $MTBF_{Ch}$. That is,

$$FR_{Ch} = \frac{1}{MTBF_{Ch}}$$

The values of FR_{Ch} are usually rather small ($\ll 1$) for on-chip interconnect, as listed in Table 4.1 of [9]. Assuming that two channels are attached between neighboring routers, and there are totally $2n$ channels in the network. Since the conventional UNI-directional NoC (UNI-NoC) fails in case of any faulty channel, the failure rate can be represented as:

$$FR_{UNI-NoC} = 1 - (1 - FR_{Ch})^{2n} \cong 2n \times FR_{Ch}$$

With bidirectional channels, if there is a pair of such channels between each pair of routers, then the NoC traffic will not be severely disrupted unless both channels fail. Therefore, the failure rate of BFT-NoC may be estimated as:

$$FR_{BFT-NoC} = 1 - (1 - FR_{Ch}^2)^n \cong n \times FR_{Ch}^2 \cong \frac{FR_{Ch}}{2} \times FR_{UNI-NoC}$$

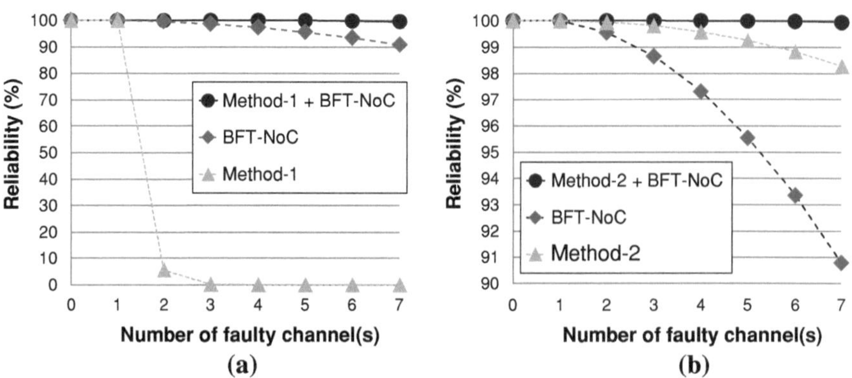

Fig. 8.8 **a** Reliabilities of Method-1, BFT-NoC, and Method-1+BFT-NoC, and **b** reliabilities of Method-2, BFT-NoC, and Method-2+BFT-NoC

Since $FR_{Ch} \ll 1$, one has $FR_{BFT\text{-}NoC} \ll FR_{UNI\text{-}NoC}$. Clearly, with bidirectional channels, the failure rate of a BFT-NoC is significantly reduced.

8.3.7 Reliability Enhancement

Once faulty channels exist in an NoC, using fault-tolerant schemes will provide a certain reliability to maintain the functional correctness of a system operation. Here, we evaluate the reliabilities of our proposed BFT-NoC and the other two schemes, proposed in [2] and [3] and respectively called as Method-1 and Method-2 in later sections. Each of the reliability data as shown in Fig. 8.8 was calculated with an 8×8 mesh. Though BFT-NoC cannot tolerate the case where channels between neighboring nodes are both faulty, but BFT-NoC provided a 90.80% reliability even if seven faulty channels existed in the network. Referring to Fig. 8.8a, Method-1 provided a poor reliability when the number of faulty channels was bigger than 1. The reason is that Method-1 sets a constraint that all faults must be related to one router. That is, if there are two faulty channels connected to four routers, deadlocks could happen when using Method-1. As shown in Fig. 8.8b, Method-2 supported a good reliability (>98%) in case of seven faults. However, Method-2 needed to re-configure the global routing tables; thus, Method-2 is difficult to dynamically handle *transient faults*. Last and most importantly, BFT-NoC can be combined with Method-1 or Method-2 as a hybrid scheme of Method-1+BFT-NoC or Method-2+BFT-NoC, which respectively improved the reliabilities to 99.58% and 99.94% as shown in Figs. 8.8a, b.

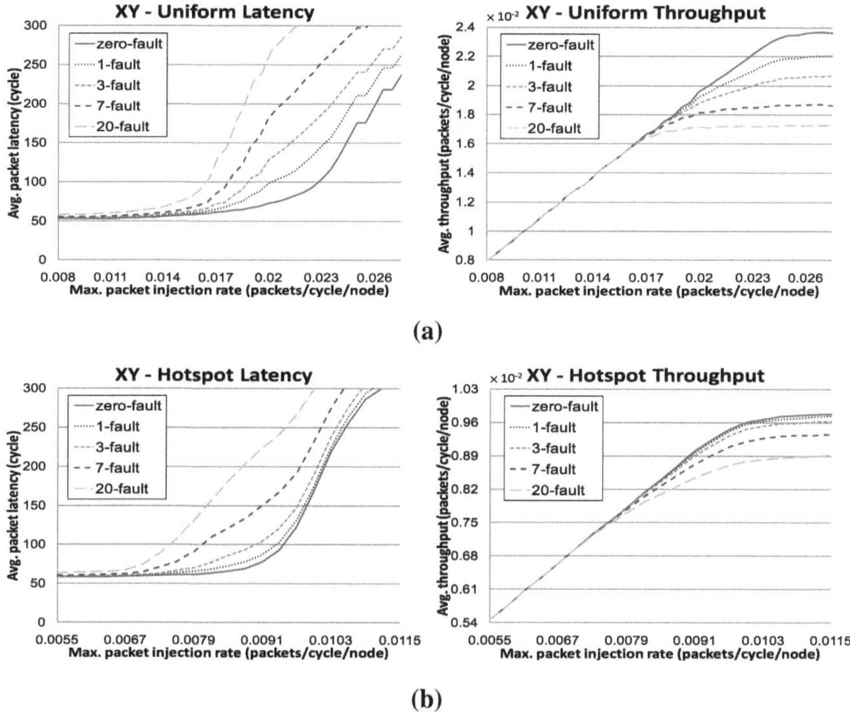

Fig. 8.9 Performance variations with XY routing algorithm under **a** uniform traffic and **b** hotspot traffic

8.4 Experimental Results

To evaluate the performance impact among various traffic types upon different numbers of faulty channels in the BFT-NoC, comprehensive simulations were run in Register Transfer Level (RTL) using Cadence NC-Verilog. Each channel bandwidth was set to one flit (32 bits) per cycle. Four cycles were required for switching a header flit to an output port in a pipelining fashion. We assigned to each input port buffer 1024 bits as in [11, 12] and applied the wormhole switching technique.

8.4.1 Experiments with Synthetic Traffics

Synthetic traffic performance analyses in terms of latency and throughput were carried out on an 8×8 mesh network. The sizes of packets were randomly distributed between 4 and 16 flits. All performance metrics were averaged over 30,000 packets after a warm-up session of 10,000 arrival packets. In uniform

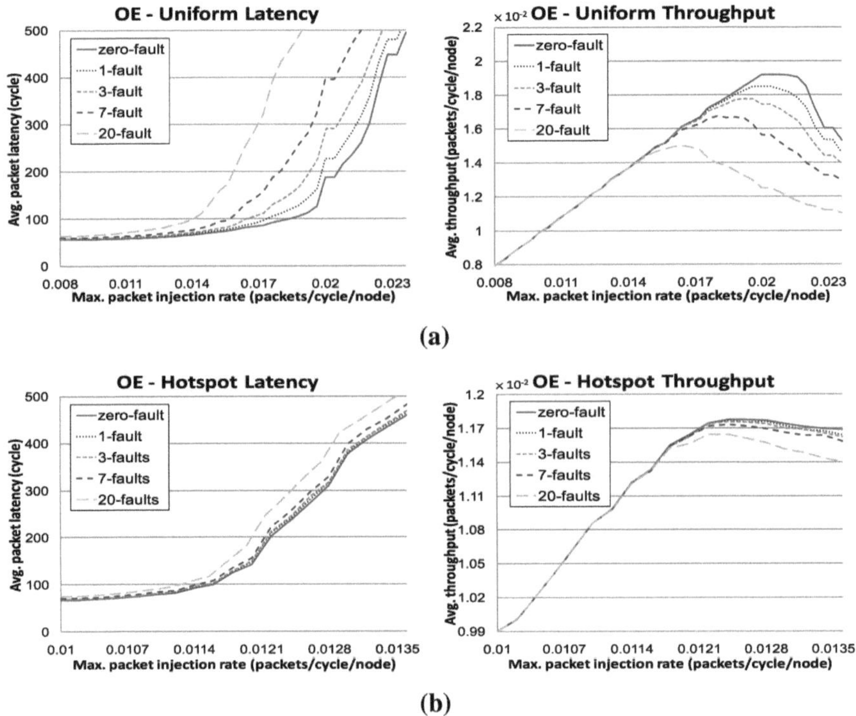

Fig. 8.10 Performance variations with Odd–Even routing algorithm under **a** uniform traffic and **b** hotspot traffic

traffic, a node transmits a packet to any other node with equal probability. In hotspot traffic, uniform traffic is applied, but 20% of packets change their destination to one of the following four selected nodes [(7, 2) (7, 3) (7, 4), (7, 5)] with equal probability. Five different faulty-channel numbers including 0, 1, 3, 7, and 20 were performed to realize the performance impacts. Each data of the performance metrics was an average value of 100 simulations with randomly selected faulty-channel locations.

As shown in Figs. 8.9 and 8.10, first, the latency increased and the throughput decreased along with the increased number of faulty channels in all simulations. Next, the performance downgrading was moderate. For example, compared with the simulations with zero-fault, the decreases of the maximal throughput in the simulations with the 7-fault were small with 20.83, 4.33, 12.82, and 0.37% in XY-Uniform, XY-Hotspot, OE-Uniform, and OE-Hotspot, respectively. Lastly, BFT-NoC performed better with Odd–Even than with XY, the reason is that the adaptive routing algorithm of Odd–Even can reduce the chance of packets going through the bidirectional channels where TX and RX data flows share with a channel bandwidth.

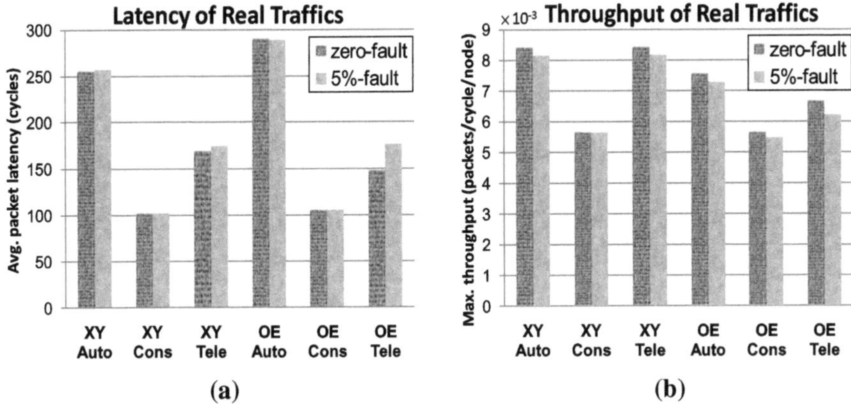

Fig. 8.11 Performance comparisons of **a** latency and **b** throughput under real traffics

Table 8.1 Implementation overhead analyses

Item \ Scheme	UNI-NoC	BFT-NoC	Overhead (%)
Router area	407762.00	424396.00	4.07
Router power	79.57 mW	81.05 mW	1.86
Router timing	1.23 ns	1.23 ns[a]	0.00

Note[a] : The 0.29 ns latency of the additional control is not located in the critical path of the designed router

8.4.2 Experiments with Real Traffics

In addition to evaluating the BFT-NoC performance with synthetic traffics, we used E3S benchmarks [13] to demonstrate the performance variations in real traffics. Experiments used the same setting for synthetic traffics, but each of the three adopted real traffic patterns: auto-indust, consumer, and telecom, was repeated 100 times and run on a 5×5, 4×4, and 6×6 mesh networks, respectively. We compared zero-fault and 5%-fault, and the experimental results are shown in Fig. 8.11.

To sum up, the performance impacts of the average latency and of the maximal throughput were 3.84% and 3.37%, respectively. The downgrading was quite small compared to that in synthetic traffics since the 5%-fault corresponds to 11 faulty channels in an 8×8 mesh network. In the analyses, we found that most data flows of real traffics were unidirectional in-between routers. In other words, even in normal operations (i.e., zero-fault) between neighboring routers, only one channel was in use for TX or RX, and the other channel was always idle. Thus, although in the fault-tolerant operation modes (i.e., 5%-fault), the bandwidth sharing for both TX and RX rarely happened in contrast to the sharing in synthetic traffics.

8.4.3 Implementation Overhead

Our router was designed in Verilog, synthesized by Synopsys Design Compiler, and power analyses were calculated by Synopsys Power Compiler in UMC 90 nm technology. Referring to Table 8.1, the area and power overheads for implementing a BFT-NoC router were small in 4.07 and 1.86%, respectively. Moreover, BFT-NoC made no timing impact since the additional controls for bidirectional channels can execute in parallel with the primary UNI-NoC design.

8.5 Remarks

In this chapter, we proposed a novel NoC fault-tolerant scheme, named Bi-directional Fault-Tolerant NoC (BFT-NoC). To the best of our knowledge, BFT-NoC is the first to utilize bidirectional channels to provide NoC with a fault-tolerance capability instead of detouring packets as in traditional schemes. BFT-NoC can dynamically deal with either *permanent faults* or *transient faults*, both of which exist in the next generation of NoCs. In the analyses, BFT-NoC is not only a stand-alone and highly-reliable (90.80%) fault-tolerant scheme, but also can be combined with other detour-based algorithms to achieve an ideal reliability of 99.94%.

References

1. T. Schonwald, J. Zimmermann, O. Bringmann, and W. Rosentiel, "Fully Adaptive Fault-Tolerant Routing Algorithm for Network-on-Chip Architectures," *in Proceedings of the Euromicro Conference on Digital System Design*, pp. 527–534, August 2007
2. Z. Zhen, A. Greiner, and S. Taktak, "A Reconfigurable Routing Algorithm for a Fault-tolerant 2D-Mesh Network-on-Chip," *in Proceedings of the Design Automation Conference*, pp. 441–446, June 2008
3. D. Fick, A. DeOrio, G. Chen, V. Bertacco, D. Sylvester, and D. Blaauw, "A Highly Resilient Routing Algorithm for Fault-Tolerant NoCs," *in Proceedings of the Design, Automation and Test in Europe Conference and Exhibition*, pp. 21–26, April 2009
4. M. Valinataj, S. Mohammadi, J. Plosila, and P. Liljeberg, "A Fault-Tolerant and Congestion-Aware Routing Algorithm for Networks-on-Chip," *in Proceedings of the IEEE International Symposium on Design and Diagnostics of Electronic Circuits and Systems*, pp. 139–144, April 2010
5. A. Kohler, G. Schley, and M. Radetzki, "Fault Tolerant Network on Chip Switching With Graceful Performance Degradation," *IEEE Transactions on Computer-Aided Design of Integrated Circuits and Systems*, vol. 29, no. 6, pp. 883–896, June 2010
6. M. H. Cho, M. Lis, K. S. Shim, M. Kinsy, T. Wen, and S. Devadas, "Oblivious Routing in On-Chip Bandwidth-Adaptive Networks," *in Proceedings of the Parallel Architectures and Compilation Techniques*, pp. 181–190, September 2009

7. Y. C. Lan, S. H. Lo, Y. C. Lin, Y. H. Hu, and S. J. Chen, "BiNoC: A Bidirectional NoC Architecture with Dynamic Self-Reconfigurable Channel," *in Proceedings of the International Symposium on Network-on-Chip*, pp. 266–275, May 2009
8. L. Benini and G. DeMicheli, "Networks on Chips: a New SoC Paradigm," *IEEE Transactions on Computers*, vol. 35, no. 4, pp. 70–78, January 2002
9. G. DeMicheli and L. Benini, *Networks on Chips: Technology and Tools*, Morgan Kaufmann, 2006
10. G. M. Chiu, "The Odd-Even Turn Model for Adaptive Routing," *IEEE Transactions on Parallel and Distributed Systems*, vol. 11, no. 7, pp. 729–738, July 2000
11. G. Michelogiannakis, D. Sanchez, W. J. Dally, and C. Kozyrakis, "Evaluating Bufferless Flow Control for On-Chip Networks," *in Proceedings of the International Symposium on Networks-on-Chip*, pp. 9–16, May 2010
12. S. R. Vangal, J. Howard, G. Ruhl, S. Dighe, H. Wilson, J. Tschanz, D. Finan, A. Singh, T. Jacob, S. Jain, V. Erraguntla, C. Roberts, Y. Hoskote, N. Borkar, and S. Borkar, "An 80-Tile Sub-100-W TeraFLOPS Processor in 65-nm CMOS," *IEEE Transactions on Solid-State Circuits*, vol. 43, no.1, pp. 29–41, January 2008
13. R. Dick, "Embedded System Synthesis Benchmark Suites (E3S)," http://ziyang.eecs.umich.edu/~dickrp/e3s/, Accessed January 2011

Chapter 9
Energy-Aware Application Mapping for BiNoC

Power-efficient scheduling is investigated for BiNoC architecture in this chapter. To minimize the power consumption of real time applications on BiNoC, time slacks in a preliminary schedule are exploited to conserve power. In addition to the processing units, wide variance in the link utilization of an NoC also leads to huge power saving if the link frequency can be tuned accurately to track the variations in bandwidth requirements. This can be accomplished by utilizing the DVS technique to scale the link voltage or frequency, as long as the deadline is met. An efficient power aware task and communication scheduling algorithm is proposed with a unique feature of utilizing the configurability of a bidirectional channel to trade the data transmission time for power expenditure. Extensive simulations are performed to compare the proposed algorithm against conventional Earliest-Deadline First (EDF) based algorithm on NoC.

9.1 Preliminaries

Given an application task graph and a BiNoC architecture, this chapter develops a novel power-efficient scheduling algorithm for task and communication considering the benefit of bidirectional link. In this algorithm, a hybrid power optimization refinement process, which can adjust link voltage to scale data transmission frequency while observing application bandwidth requirement, is proposed. In summary, we want to find (1) a task-to-processor mapping that determines which processor the task is executed on, (2) a schedule of tasks such that all the hard deadline constraints are met, and (3) a communication event schedule which can avoid traffic congestion and maximize the benefit of the BiNoC architecture while optimizing power consumption.

S.-J. Chen et al., *Reconfigurable Networks-on-Chip*,
DOI: 10.1007/978-1-4419-9341-0_9,

9.1.1 Task and Communication Scheduling

In those previous works introduced in Chap. 5, only a few of them considered the effect caused by communication between tasks. Varatkar and Marculescu [1] proposed a design flow with some consideration of communication, but they only used a proportion factor to measure the impact of communication without any model. Hu and Marculescu [2] proposed an algorithm that considers the communication model of an NoC architecture; however, they use the distance model of an NoC architecture to calculate the communication energy, not for the optimization of the system performance. Most important of all, so far there have been no task scheduling algorithms that consider the communication model of a BiNoC architecture as described in this chapter.

Compared with the communication architecture such as a Bus or a conventional NoC architecture, BiNoC architecture has more flexibility in data transmission. For example in a conventional NoC, only one channel on the routing path will be used when routing path is decided. The reason is that the transmission direction of the two channels between two routers is fixed, and only one of the two channels has the same direction as the routing path. However, the direction of the channels can be changed in a BiNoC architecture. That is, if another channel is not used at the time, both channels can be used for data transmission. The transmission time of the same data amount could be reduced to half of the original time required in a conventional NoC architecture. If the transmission time can be reduced on a many-core system, the overall execution time can also be reduced.

Although the BiNoC architecture has the benefit that can shorten the transmission time to half, sometimes this advantage may not play because of the feature of a mesh-based architecture. An example is shown in Fig. 9.1. There are two data transmissions on the network at the same time and the routing path of these transmissions overlap. These two transmissions can exist concurrently with no contention because of the advantage of a BiNoC architecture which can change the direction of channels. However, either of these two transmissions cannot shorten the transmission time, because they can only get the usability of one channel at the bottleneck spot. Therefore, if we want to fully utilize the transmission benefit of a BiNoC architecture, the traffic of the routing path needs to be known in advance such that the usability of both the two channels on the whole routing path is guaranteed. Therefore, communication scheduling that can arrange the transmission and avoid the congestion is very important for the BiNoC architecture.

9.1.2 Communication Model of BiNoC Architecture

A BiNoC architecture is generally specified as a undirected graph $A(P, CL)$ to represent the set of processing elements (PE) and routers that are connected by a specific network topology represented by the channel links (CL) in the platform.

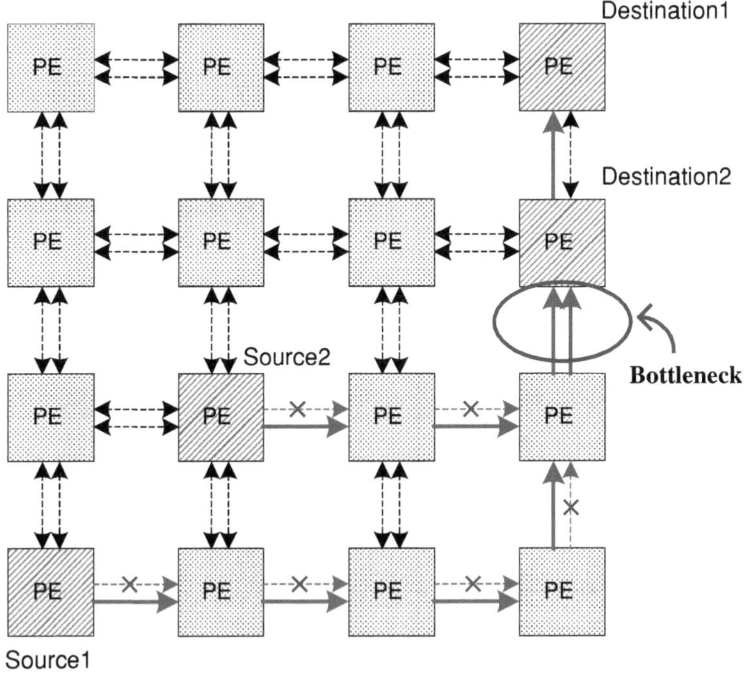

Fig. 9.1 Examples of traffic on BiNoC architecture

Each vertex $p_i \in P$ denotes a processing element that is annotated with relevant information with respect to the type of a processor. Each undirected line $l_{i,j} \in CL$ represents a data transmission channel between p_i and p_j, which is also associated with information such as data bandwidth $bw(l_{i,j})$.

In the BiNoC architecture graph $A(P, CL)$, each vertex p_i is connected with its neighbor vertex by two undirected lines $l_{i,j}$ and $l_{j,i}$. Undirected line $l_{i,j}$ represents that the communication request from p_i has the higher priority, also it is opposite for line $l_{j,i}$. This priority design is for that when both p_i and p_j want to transfer data, each side can get grant of one channel to transfer. This also maintains the compatibility of the characteristics of conventional NoC, and prevents the problem of starvation.

Given a task graph and a BiNoC architecture, the output of our scheduling algorithm is a task schedule and a communication schedule that suit for BiNoC architecture. The target is:

$$minimize\left\{\mathrm{Max}_{i=1}^{N}[\text{finish time }(\tau_i)]\right\}$$

such that

- All the task dependences are compliant.

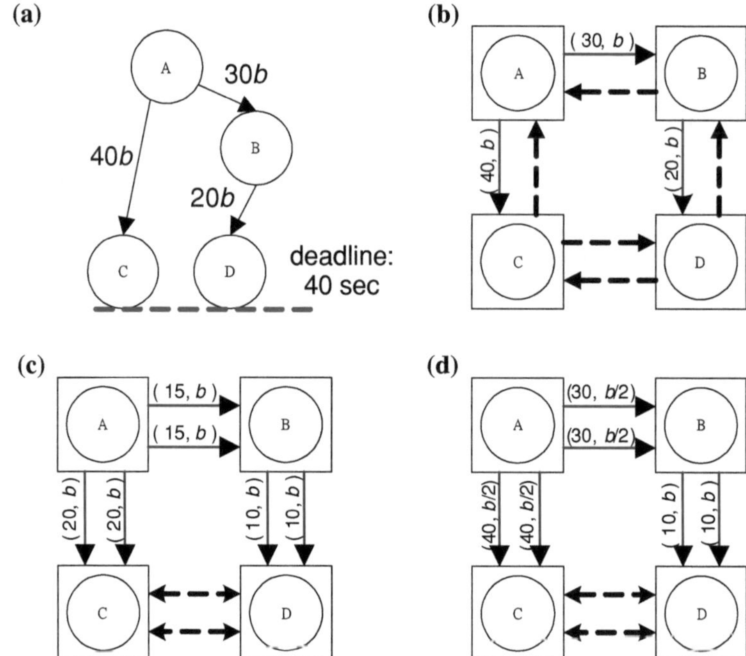

Fig. 9.2 Example of task graph mapping to T-NoC and BiNoC

- All the task computations on the same PE assure mutual exclusion policy.
- All the communications on the same channel assure mutual exclusion policy.

9.2 Motivational Example

To illustrate the benefit of choosing BiNoC as backbone architecture and the potential power-saving capability of BiNoC, an elaborate example in Fig. 9.2 is given as follows.

Figure 9.2a illustrates an application task graph with deadline requirement of 40 s and each edge represents the communication dependence with a value of communication volume (*bits*). For easier comparison, task execution times are ignored in this example. An optimized mapping of this application onto a 2 $\times$ 2 Typical-NoC is shown in Fig. 9.2b, where only three channels are used during the whole simulation. The label (t, b) on a link represents that the data needs to be transmitted for duration t (*sec*) with bandwidth b (*bits per sec*). Since the critical path is from tasks A to D through task B, the total data transmission time here is 50 s which violates the deadline requirement. Considering the same example on BiNoC as illustrated in Fig. 9.2c, deadline requirement can be met since data

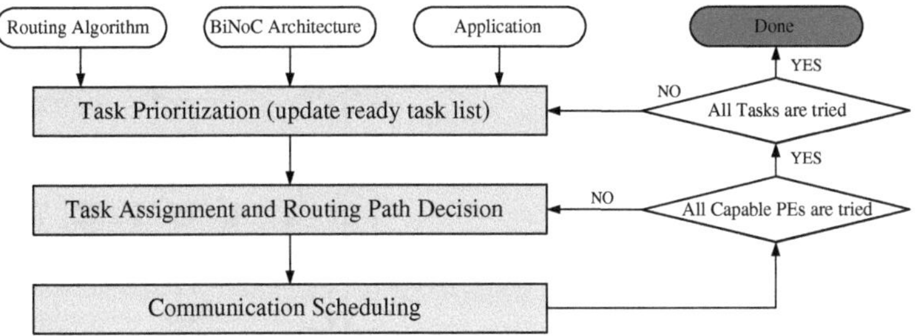

Fig. 9.3 EDF based task and communication scheduling

transmission can be shared by the idled link, which improves the total transmission time to 25 s. However, there will be respectively a 20 s and a 15 s time slacks after the ends of tasks *C* and *D*. These time slacks can be exploited to reduce link power consumption by regulating the supply voltage to a desired link frequency. As illustrated in Fig. 9.2d, by scaling down the voltages on links *AC* and *AB*, transmission bandwidth is reduced from *b* to *b*/2 thus all the time slacks can be utilized for power optimization without deadline violation. In other words, by leveraging the advantages of bidirectional channel in BiNoC, we can further improve the power consumption by dividing data into idled channel and scaling down the link operating frequency (bandwidth) without increasing the latency. Take the data transmission between tasks *A* and *C* for example, power consumption is decreased according to the *b*/2 bandwidth even the total transmission time is still 40 s in Fig. 9.2d. Therefore, we are motivated to apply DVS technique to link, in order to provide variable transmission bandwidth and thus optimize the BiNoC interconnection network power expenditure.

9.3 Task and Communication Scheduling for BiNoC

Earliest Deadline First (EDF) technique is used as baseline of our scheduling algorithm, which has been proven to achieve the optimal performance result on single-processor system constructively. For the many-core system, the simplicity of the EDF algorithm makes the inclusion of communication consideration and refinement process much easier. The flow of our baseline task and communication scheduling algorithm is illustrated in Fig. 9.3.

As a list scheduling based algorithm, EDF prioritizes the tasks which predecessors are all scheduled according to their deadlines and puts them into a ready task list (RL) iteratively. Then, the object of task assignment is to determine a PE and its corresponding routing path for the task, selected at the task prioritization step. Thereafter, given the information of routing path and configurability characteristic of a BiNoC, the exact data transmission time has to be set to avoid congestion in the communication scheduling step. After all the candidate PE's on

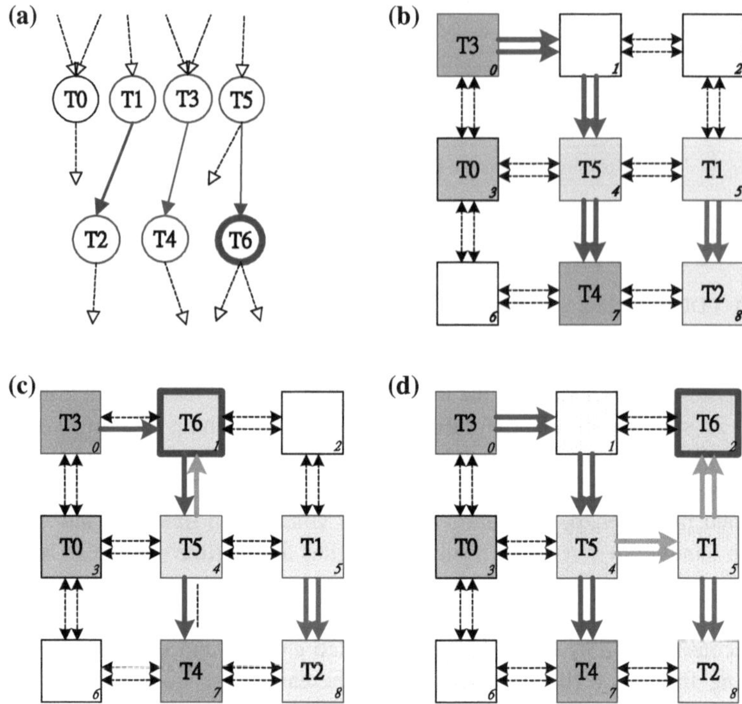

Fig. 9.4 Communication scheduling under BiNoC consideration

the system have been tried, we can determine which location will be assigned to a task. Then, while all the tasks are scheduled, a performance oriented scheduling result is generated.

Figures 9.4a, b respectively show a sub-section of a task graph and its corresponding assignment. We can find that different locations of task *T*6 will affect the communication efficiency of the rest of the tasks on BiNoC as illustrated in Figs. 9.4c, d. In this case, the assignment of *T*6 as shown in Fig. 9.4d could be more efficient even if its location is farther away from *T*5, since BiNoC could provide more available transmission bandwidth.

9.3.1 Communication Model and Traffic on BiNoC

In this section, we will use some examples to illustrate the difference in congestion between the conventional NoC architecture and the BiNoC architecture, and the way to solve congestion for the BiNoC architecture.

Figure 9.5a shows the transmission congestion on the conventional NoC architecture. Since the channels on the conventional NoC architecture are

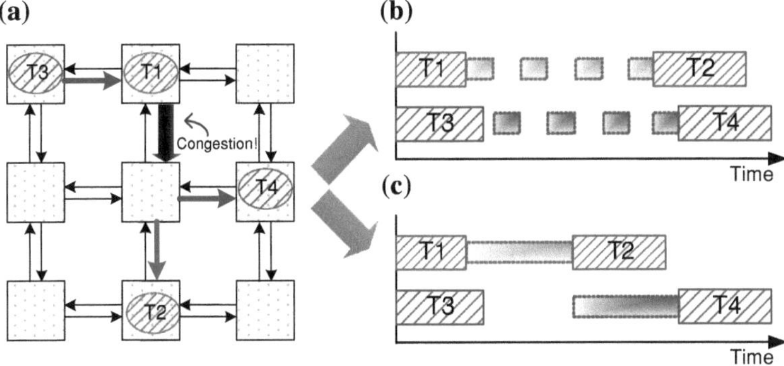

Fig. 9.5 Network congestion and solutions on NoC architecture

unidirectional and unchangeable, there are two possible solutions when two data transmissions want to use the same channel: One is shown in Fig. 9.5b, where the two data transmissions take turns using the channel with a certain time interval. Another is shown in Fig. 9.5c, where the first data transmission gets to use the whole channel, the second data transmission cannot start until the first data transmission is complete. The former method seems much fair than the latter, but actually the latter method is better for performance. It is clear that task T4 starts execution at the same time in both two methods, but task T2 can start earlier by using the latter method. According to this example, we can see that when the congestion occurs, the better way of using a NoC architecture to solve this problem is allowing the first data transmission to use the whole channel. This is the better way in term of performance.

The same situation will happen on the BiNoC architecture when both data transmissions want to use the two channels to transfer data, but the transmission time can be cut in half. Both channels will encounter the same congestion. This situation is very much like the congestion on the conventional NoC architecture, so the appropriate solution is also the same as on the conventional NoC architecture. Figure 9.6 shows the transmission congestion and the solutions on the BiNoC architecture.

So what is the difference in traffic congestion between the conventional NoC architecture and the BiNoC architecture? Although the data transmission on the BiNoC architecture can use both of the channels to transfer data, it is not necessary to and does not always start and terminate the transmission at the same time on both of the channels. Sometimes the congestion occurrences on both channels will not be at the same time or only one channel has congestion. There is an example shown in Fig. 9.7. For some reason, the data transmission between tasks T1 and T2 only uses one channel to transfer, and the data transmission between tasks T3 and T4 wants to use both channels to transfer. To solve this congestion, there are two weak solutions which are shown in Fig. 9.7b, c. One solution shown in Fig. 9.7b is to delay both transmissions on the two channels. This is not a good solution

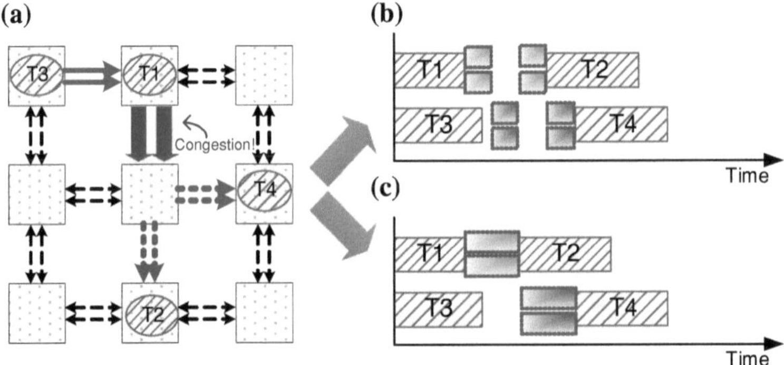

Fig. 9.6 Network congestion and solutions on BiNoC

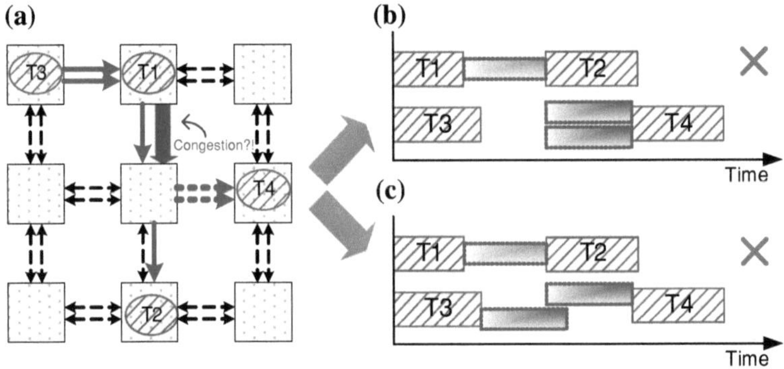

Fig. 9.7 Single channel congestion on BiNoC

obviously because there is one unused channel at the earliest time that the transmission can start. Another solution as shown in Fig. 9.7b is to only delay the transmission on the congested channel. This is also not a good solution because the bottleneck is still the delayed channel, and the execution start time of task T4 is the same as the first solution. Hence, both two solutions are not good solutions and both of them do not exploit the advantages of the BiNoC architecture to the extreme.

In fact, task T4 can start execution earlier by properly arranging the data transmission on the BiNoC architecture, such that the total performance of the application can be better. This better transmission method is shown in Fig. 9.8. The amount of data transmission on each channel should be adjusted. We can use the channel without congestion to transfer more data, and delay the data on the channel with congestion to transfer less data. Hence, the end of transmission time can be earlier than the above two weak solutions. As a result, this example makes it clear again that communication scheduling is very important on the BiNoC architecture.

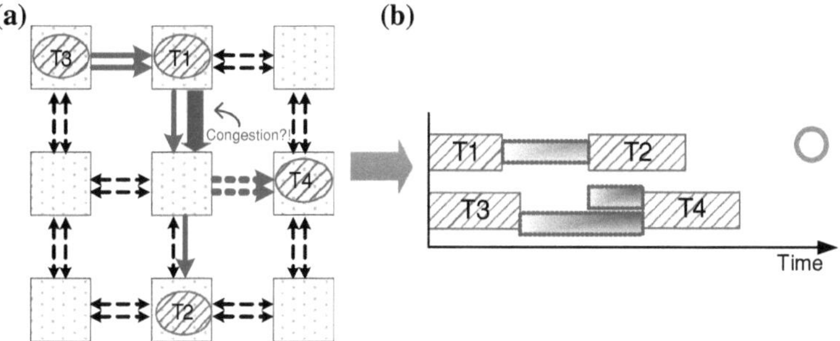

Fig. 9.8 Adventage of BiNoC to solve congestion

9.3.2 Performance Refinement Process

In this section, we will introduce the performance refinement process that we used in our algorithm. Since our proposed task and communication scheduling algorithm is a constructive algorithm, it is fast in computation but has no refinement ability when the result cannot meet the performance goal. Therefore, a separate refinement process is needed in our proposed flow to improve the performance. Since our proposed scheduling algorithm attaches importance to communication, the main purpose of our performance refinement process also is to find the communication which can be reduced or sped up. The flow of our performance refinement process is show in Fig. 9.9. As shown in Fig. 9.9, our proposed performance refinement process has many steps, and we will describe it in detail in the following sections.

The target of our proposed algorithm is to minimize the finish time of the latest task which finish time equals to the overall runtime. If this overall runtime did not reach our requirement, we want to figure out what has caused the last task finish so late. Thus, we must find the critical path of this scheduling result and optimize it. The critical path here is not just according to the execution time of the task, but also to the data transmission time on the final schedule. The discovery of the critical path starts from the latest finished task, and finds the latest data transmission of all transmissions which predecessors transfer to it. This predecessor and the data transmission now are the members on the critical path. Then the same search will be applied to that predecessor task. Finally the path which interlaced with tasks and data transmissions is the critical path that we want.

Then, we will refine the communication time of the critical path. We have found there are two cases which may reduce the data transmission time on the path. The *Case* 1 & *Case* 2 *Searching* will be introduced in detail with examples as follows.

Figure 9.10 shows an example of Case 1. The task graph is shown in Fig. 9.10a, which critical path is T1 → T2 → T4. The original scheduling result is shown in Fig. 9.10b. Case 1 is that there is a task on the critical path, and this task has only

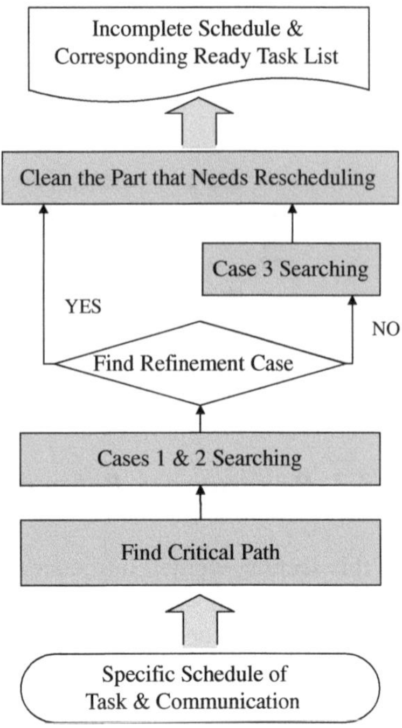

Fig. 9.9 Flow of our performance refinement process

one predecessor. Also, the PE to which the task is allocated and the PE to which its predecessor is allocated are not the same PE. In this example, task T2 matches the condition of Case 1. Task T2 is on the critical path. Task T2 only has one predecessor which is task T1, also T2 and T1 are allocated to different PEs. For the task which matches the condition of Case 1, the refinement is to reallocate the task to the PE which predecessor is executed on. Therefore, the data transmission between them can be eliminated. Besides, the other data transmission on the critical path will use two channels to transfer forcibly. For this example, the result of Case 1 refinement is shown in Fig. 9.10c. We can see that because of the elimination of data transmission between tasks T1 and T2, the end time of task T4 which equals to the overall time is earlier than the original schedule.

Figure 9.11 shows an example of Case 2. The task graph is shown in Fig. 9.11a, which critical path is T1 → T2 → T4. The original scheduling result is shown in Fig. 9.11b, same as the example in Case 1. Case 2 is that there is a task on the critical path, and this task has more than one predecessor. However, the PE to which the task is allocated is different from all the PEs to which its predecessors were allocated. In this example, task T4 matches the condition of Case 2. Task T4 is on the critical path and has two predecessors which are task T2 and T3, also T4 is allocated to the PE which is different to the PEs to which T2 and T3 were allocated. For the task which matches the condition of Case 2, the refinement is to reallocate the task to the PE on which the most critical predecessor was executed. Therefore, the data transmission between them can be eliminated. Besides, the

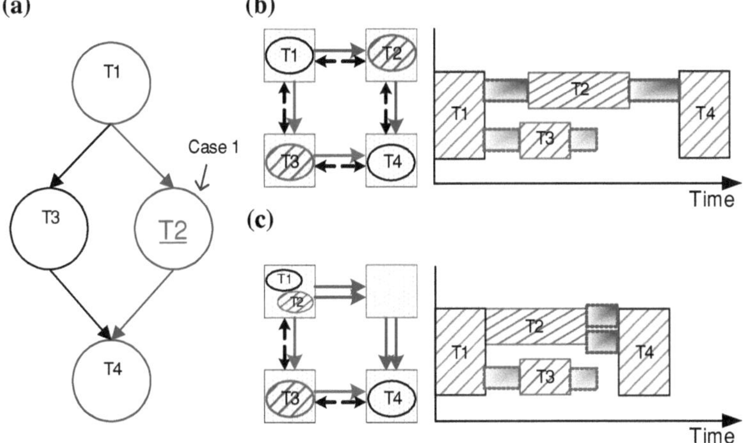

Fig. 9.10 Example of Case 1 refinement

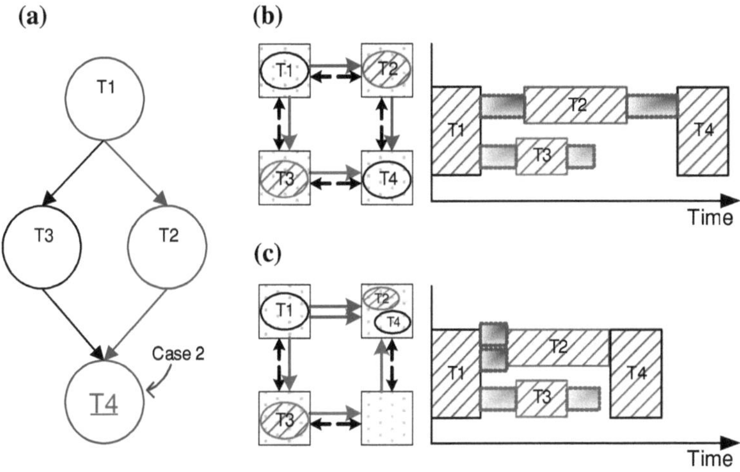

Fig. 9.11 Example of Case 2 refinement

other data transmission on the critical path will use two channels to transfer forcibly. For this example, the result of Case 2 refinement is shown in Fig. 9.11c. We can see that because of the elimination of data transmission between tasks T2 and T4, the end time of task T4 which equals to the overall time is earlier than the original schedule.

About the specific method of schedule refinement, we do not move or fix it individually. The reason is that the change of one task may cause a big difference on the execution times of tasks that will be executed after the change, and the better choice of task allocation may become different too. Also, the refinement of

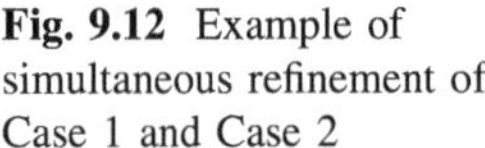

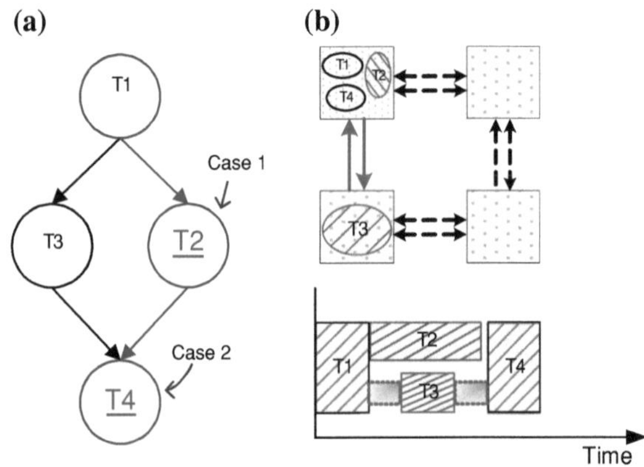

Fig. 9.12 Example of simultaneous refinement of Case 1 and Case 2

cases that we found at the previous step may have some connection to each other. Taking the example at the previous stage for instance, the best schedule result is the simultaneous implementation of both Case 1 and Case 2 refinements as shown in Fig. 9.12. It is the combination of the Case 1 (moving task T2 with task T1) and Case 2 (moving task T4 with task T2).

As stated in the reasons above, the method of schedule refinement applied in our refinement flow is to set critical flag to the tasks that were found by the case searching step. Then we put the refined result to our task and communication scheduling process to perform rescheduling. When performing rescheduling, the tasks which have set the critical flag will be chosen first at the "Choose the Highest Priority Task" step even if it has a later deadline. At the "PE Assignment" step the task also will be allocated to the PE to which its predecessor was allocated. Besides, since we use rescheduling instead of moving and fixing tasks, the schedule for the refined tasks will be adjusted to the best result automatically. Therefore, we can get a correct and quite optimized schedule from the small change of tasks.

It is worth mentioning that although there is quite a good chance to improve the performance by using Case 1 and Case 2, there is no guarantee for improvement. Sometimes the schedule obtained after the refinement process will be worse than the original schedule. Therefore, when such case is found, the refinement process will decide whether the case can be accepted or not by a sieve function. The details will be introduced in the next section.

If no case was found at the *Case* 3 *Searching* step or all the cases which were found are not accepted by the sieve function, this step will be applied to find a new possibility for performance improvement. The task on the critical path will be set the critical flag randomly, such that these tasks will get higher priority while performing rescheduling. Thus, the new schedule will be different from the original. By collocation with the sieve function, we expect this random *Case* 3 *Assignment* can achieve similar effect as GA or SA.

As mentioned in the step of Case Searching, we applied rescheduling to adjust the schedule altered by task moving. Hence, the schedule after the task moving must be cleaned for rescheduling. However, there are some schedules of tasks that may not need to be cleaned. This is because our scheduling algorithm is a constructive algorithm and these tasks are allocated and scheduled much earlier than the first task which wants to change. Thus, in the step of *Clean the Part that Needs Rescheduling*, not the whole schedule needs to be cleaned and rescheduled. In practice, we record the order of tasks that are selected by the choosing step, and preserve the schedule from the first task to the last predecessor of the first task which needs to be changed. The schedule after the last predecessor of the first task will be deleted. Then we scan the whole task list to recreate the ready task list. With this ready task list and the incomplete schedule, we can perform the rescheduling faster than the whole rescheduling on the total task graph.

9.3.3 Self-Study and Sieve Framework

Since the elimination of one data transmission will cause the change of critical path to another path and make the overall runtime longer, we want to construct a *sieve framework*, such that the refinement process knows which case has a good or a bad performance.

When a refinement process is performed for the first time, we do not actually know which case is good for performance. Therefore, we used a *self-study framework* that can evolve itself by repeatedly performing the refinement process. The flow of our self-study and sieve frameworks is shown in Fig. 9.13. We give every task a value called "Refinement Threshold," which will be changed by the result of the refinement process. Every time we find a case for refinement, we will randomly choose a number and compare it with the "Refinement Threshold" of the task. If the "Refinement Threshold" of the task is higher than that a random number, we will accept this case. Otherwise, the case will be refined this time.

In other words, the *Refinement Threshold* of the task represents the possibility of a task being refined. If the *Refinement Threshold* of the task is high, it means that performing refinement of this task has high possibility to make the total performance better. Otherwise if the *Refinement Threshold* of the task is lower, it means that performing refinement of this task may cause a negative impact on the total performance. Therefore, if we adjust the *Refinement Threshold* of tasks by the results of refinement process every time, it will learn to distinguish good cases and bad cases. Consequently, this *Refinement Threshold* of tasks is the key to our self study and sieve framework.

There is a range of *Refinement Threshold*, we use 1–100 in this framework. We initialize the *Refinement Threshold* of every task to 50 at the beginning. Then, when a task is found by the Case Searching step the first time, the possibilities of acceptance and rejection are the same. However, with the subsequent increase in the number of refinement processes, the *Refinement Threshold* of tasks which can

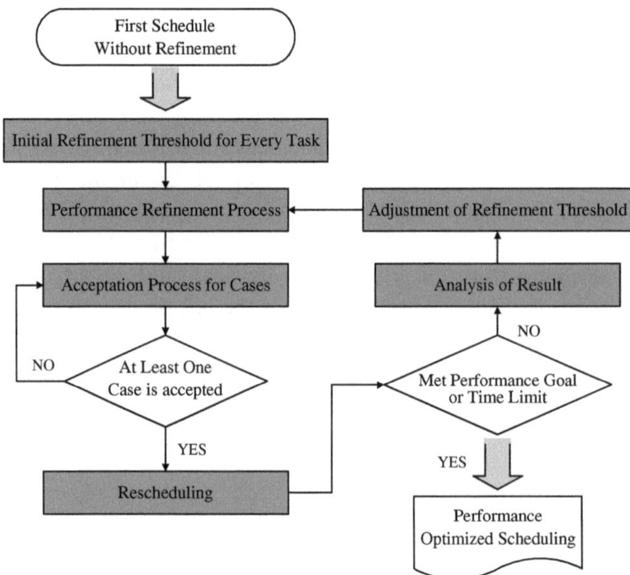

Fig. 9.13 Flow of our self-study and sieve framework

make performance better will become bigger while the *Refinement Threshold* of tasks which can make performance worse will become smaller. The specific way used to adjust the value of the *Refinement Threshold* is shown below:

1. If the result of rescheduling at the current time is the best result ever, then we substantially increase the *Refinement Threshold* of the tasks accepted to refine at this time.
2. If the result of rescheduling at the current time is not the best but better than last time, then we slightly increase the *Refinement Threshold* of the tasks accepted to refine at this time.
3. If the result of rescheduling at the current time is worse than last time, then we slightly decrease the *Refinement Threshold* of the tasks accepted to refine at this time.

After repeating many performance refinement processes, the self study and sieve framework will lead the final schedule to a better result. For avoiding the infinite loop occurrence, we will set a performance goal and a time limit for the refinement process. When the performance goal is met or the time of refinement process reaches the limit, the whole refinement process will be terminated and output the best result obtained so far. With combination of the constructive scheduling algorithm, the fast performance refinement process, and the self study and sieve framework, we can obtain a sub-optimal result which performance is close to the best solution in a small runtime.

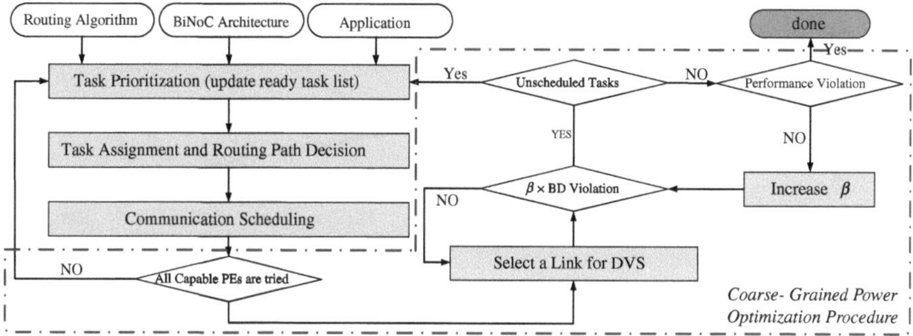

Fig. 9.14 Coarse-grained power optimization procedure

9.4 Proposed Power Optimization Solution

Our proposed power-efficient scheduling solution includes assigning the tasks of an application to suitable processors and ordering the task execution and communication sequence on each resource. For power management, we integrate DVS technique onto BiNoC that can efficiently slow down transmission frequency while keeping performance by adjusting its channel direction. This section will describe the details of our proposed power optimization procedure.

9.4.1 Coarse-Grained Power Optimization

The previous sections were focused on the performance aspects of optimization. Now, we can further improve the power consumption by using DVS technique to slow down data transmission frequency if there is any slack left to be used.

As illustrated in Fig. 9.14, the link which has slacks for power optimization will be selected after PE assignment. To decide the reference budget for each task, a budget deadline (BD) [3] will be calculated and weighted by a parameter β. By iteratively increasing the value of β while the performance (deadline) constraint is not violated, time slacks will have more opportunity to be used for power optimization as illustrated in Fig. 9.15.

9.4.2 Fine-Grained Power Optimization

Compared to the constructive method in the previous section, fine-grained power optimization recursively refines the voltage of the link which has the largest slack and re-schedules the transmission. As illustrated in Fig. 9.16, the output of EDF based task and communication scheduling is used as the initial result for power

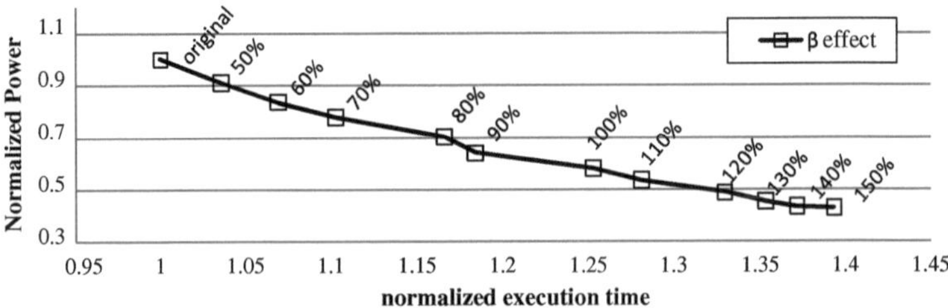

Fig. 9.15 Effect of β in coarse-grained power optimization

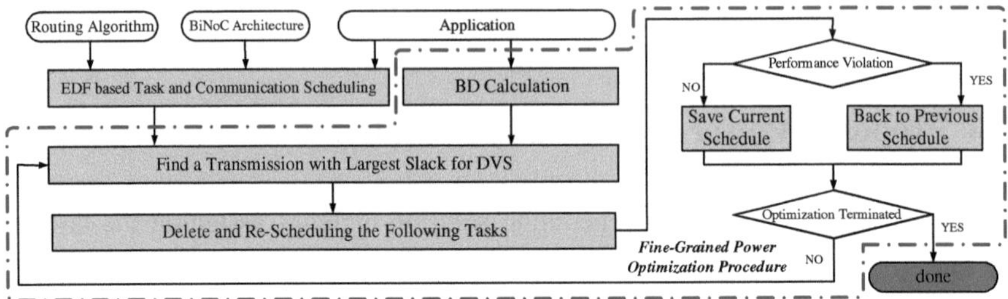

Fig. 9.16 Fine-grained power optimization procedure

optimization. The optimization will be terminated either when there is no slack for transmission or a pre-defined iteration count is reached. Note that the fine-grained power optimization will take longer refinement time than the coarse-grained method since each iterative cycle contains a partial re-schedule.

9.4.3 Proposed Power-Efficient Scheduling

To understand the features of the two above algorithms, Fig. 9.17 illustrates the power saving capability of these two methods across various deadline constraints of the same set of input task graphs. For easier comparison, power consumption was normalized to the results of EDF based scheduling. Besides, deadline constraints were extended for power optimization evaluation. The detailed experimental setup will be presented in Sect. 9.4. We can find that the fine-grained power optimization performs better when deadline constraint is tighter since this method can carefully examine all the slacks for DVS adjustment. However, as deadline constraint becomes relaxed, the coarse-grained power optimization will have more opportunity to change or slow down the critical path, thus resulting in a power optimal solution.

In order to obtain the benefits of both algorithms, as illustrated in Fig. 9.18, instead of using the result of EDF based scheduling, we use the coarse-grained power optimization result as the input to the fine-grained power optimization procedure.

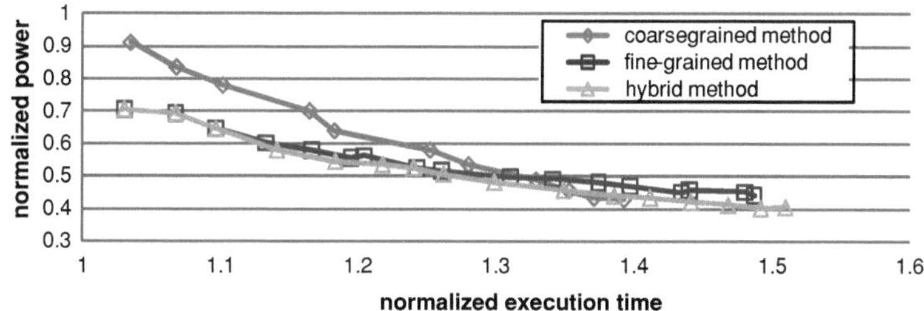

Fig. 9.17 Normalized power versus deadline constraint of two methods

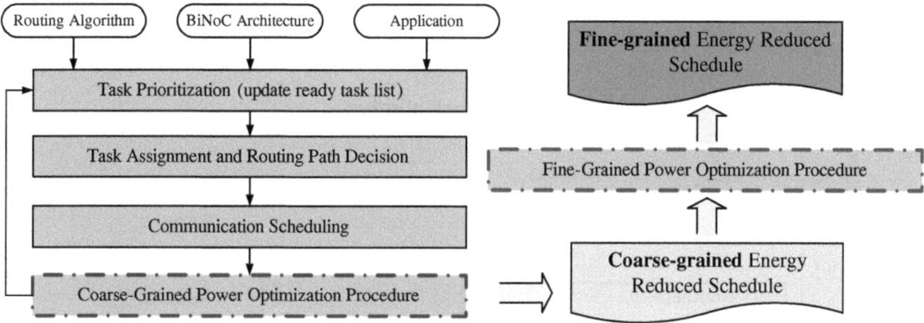

Fig. 9.18 Proposed hybrid power optimization procedure

As illustrated in Fig. 9.17, we can find that this hybrid method performs better in power consumption across various performance constraints (execution time).

9.5 Experimental Results

In this chapter, we will evaluate the effectiveness of our proposed algorithm. First, the basic experimental results will show whether the BiNoC architecture is better than the conventional NoC architecture, and the effectiveness of our task and communication scheduling algorithm for the BiNoC architecture. Secondly, the improvement of power optimization refinement process is shown by experiments.

To evaluate the effectiveness of our framework in power saving, we conducted experiments on various task graphs generated by the standard package TGFF [4]. To obtain accurate on-chip communication estimation, both NoC and BiNoC were implemented with an 8×8 mesh-based architecture for performance analysis. We also applied the dimension-ordered XY routing restriction for communication traffic. For easier comparison, all the results were normalized to EDF-based scheduling running on NoC without DVS technique applied. To simplify this problem, Table 9.1 lists the architecture with scheduling algorithms and supported voltage for comparison. In addition, since data transmission power holds a large

Table 9.1 Architectures used for experiments

Architecture	NoC	BiNoc	
Algorithm	EDF	EDF	Proposed
Supported link voltage	V	V	V,V/2

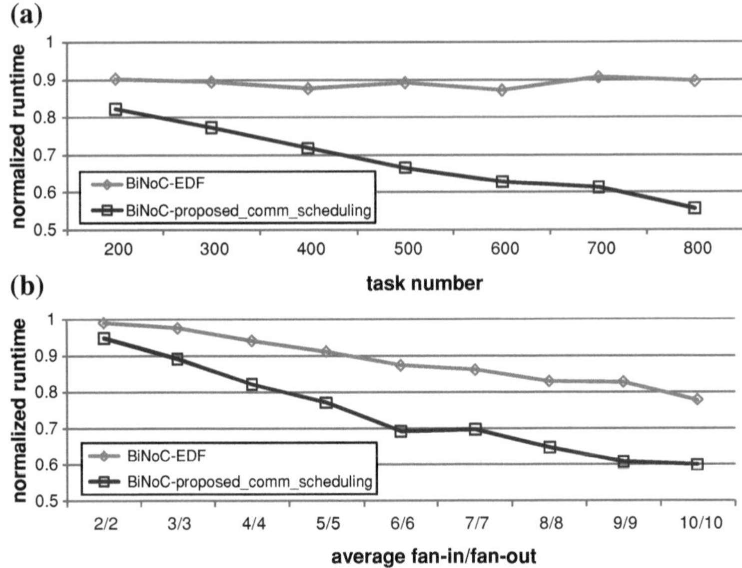

Fig. 9.19 Normalized runtime of various **a** task number and **b** complexity of communication

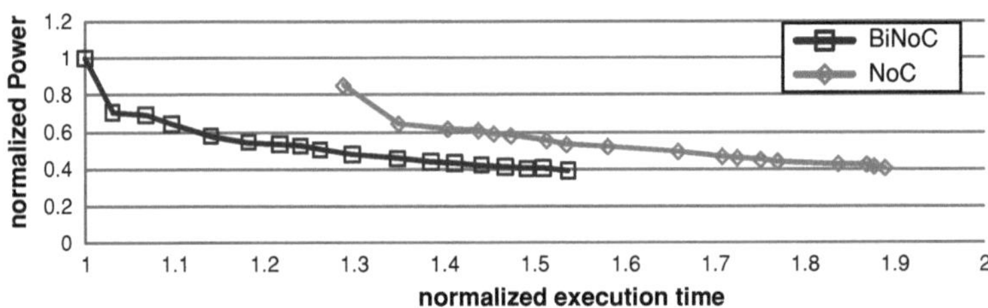

Fig. 9.20 Comparison between BiNoC and NoC architectures

portion of the total power consumption, we only focused on the power on a link in this work. Power consumption on a processing element was not taken into account.

Task graphs were generated for experiments from two manipulated variables. The effects of task number and the complexity of communication were compared as shown in Fig. 9.19. By observation of *BiNoC-EDF*, we can find that BiNoC architecture is around 10% better than conventional unidirectional NoC in performance across various numbers of tasks because of its bidirectional transmission feature. In addition, from comparisons between *BiNoC-EDF* and *BiNoC-proposed_comm_scheduling*, we can find that our proposed communication scheduling can efficiently improve the performance by utilizing the flexibility of the bidirectional links.

For the communication aspect, Fig. 9.19b uses the average fan-in/fan-out of tasks to represent the communication complexity of the whole task graph. As the number of average fan-in/fan-out becomes larger, the benefit of applying BiNoC architecture in our proposed communication scheduling was increased from 20 to 40%. Again, these results confirmed that BiNoC can efficiently decrease application run time thus has more potential to utilize these slacks for power saving by DVS technique.

Then, to compare the power saving capability of BiNoC with respect to conventional NoC architecture, Fig. 9.20 illustrates power consumption versus performance (runtime) where all the results were normalized to BiNoC with our proposed communication scheduling and power optimization algorithm under strictest deadline constraints. In this experiment, the total task number was set to be 400 and the average fan-in/fan-out was 6/6. Each point was averaged with ten task graphs generated by the TGFF tool with the above parameters. We can find that with the same power consumption, BiNoC can always finish the execution earlier than conventional NoC. On the other hand, looking at these two architectures under the same execution time (performance), power consumption on BiNoC is also lower than NoC.

9.6 Remarks

In this chapter, we proposed a power-efficient scheduling process, including task allocation and on-chip data communication ordering, by utilizing a technique of dynamic voltage scalable link. In addition, with the configurability feature of bidirectional channels on BiNoC, our proposed algorithm can efficiently trade the data transmission time for power since the direction of the idled channels can be changed to provide higher transmission bandwidth, which slows down link frequency while keeping the same performance requirement. Experimental results using TGFF standard benchmark showed that our proposed communication scheduling for BiNoC improved efficiently the data transmission time and left more slacks for power optimization. Besides, the proposed hybrid optimization algorithm indeed improved power consumption across various deadline (performance) constraints.

References

1. G. Varatkar and R. Marculescu, "Communication-Aware Task Scheduling and Voltage Selection for Total Systems Energy Minimization," in *Proceedings of the International Conference on Computer-Aided Design*, pp. 510-517, November 2003

2. J. Hu and R. Marculescu, "Energy-Aware Communication and Task Scheduling for Network-on-Chip Architectures under Real-Time Constraints," in *Proceedings of the Design, Automation and Test in Europe Conference and Exhibition*, pp. 234-239, February 2004
3. R. Mullins, A. West, and S. Moore, "Low-Latency Virtual-Channel Routers for On-Chip Networks," in *Proceedings of the International Symposium on Computer Architecture*, pp. 188-197, June 2004
4. KK. Kim, S. J. Lee, K. Lee, and H. J. Yoo, "An Arbitration Look-ahead Scheme for Reducing End-to-End Latency in Networks-on-Chip." in *Proceedings of the International Symposium on Circuits and Systems*, pp. 2357-2360, May 2005

Chapter 10
Concluding Remarks

In the first part of this book, we introduced a state-of-the-art on-chip interconnection network design and some of the important design problems of NoC. Then, in the second part of this book, many important design issues on NoC, including routing technique, reliability, and energy-aware application mapping, were described.

First, we proposed a novel BiNoC backbone architecture using dynamically reconfigurable bidirectional channels to improve bandwidth utilization with effective implementation costs. A new, distributed channel-direction control protocol that supports real-time traffic-direction arbitration while avoiding deadlock and starvation was presented. Experimental results using both synthetic traffic patterns and E3S benchmarks verified that the proposed BiNoC backbone architecture can significantly reduce the packet delivery latency at all levels of packet injection rates. Compared to the conventional NoCs, the bandwidth utilization and traffic consumption rate of our BiNoC also exhibited higher efficiency. Furthermore, it is very encouraging that the BiNoC can even achieve better latency results under a variety of traffic patterns while using less buffer size.

Based on this BiNoC backbone architecture, we implemented a QoS-aware BiNoC architecture which can arbitrate the inter-router channel direction based on real-time traffic conditions. Specifically, an inter-router channel-direction arbitration scheme which assigns a higher priority for the critical GS traffic to traverse the network was presented. Moreover, a flexible virtual-channel management mechanism and a novel prioritized routing policy were integrated in our QoS design to further enhance the communication efficiency of the GS packets.

To handle the static and dynamic channel failures on the data-link layer, a novel fault-tolerant NoC scheme using bidirectional channel, called as Bidirectional Fault-Tolerant NoC (BFT-NoC), was proposed. This mechanism is devised to mitigate potential performance hits due to faulty channels through dynamic sharing of other surviving channels. As such, costly data rerouting can be avoided, and the performance downgrading was moderate when faults existed in the network.

S.-J. Chen et al., *Reconfigurable Networks-on-Chip*,
DOI: 10.1007/978-1-4419-9341-0_10,

Furthermore, a power-efficient scheduling was presented for BiNoC to minimize the power consumption of real time applications. Specifically, time slacks in a preliminary schedule were exploited to conserve power by utilizing the DVS technique to scale the link voltage or frequency, as long as the deadline was met. The unique feature of utilizing the configurability of a bidirectional channel to trade the data transmission time for power expenditure did provide better results in BiNoC.

Compared to the conventional NoC, our proposed BiNoC backbone design did efficiently improve the performance while guaranteeing QoS and fault tolerance requirements. For power saving, the configurability of bidirectional channel design can further trade the data transmission time for power expenditure during application scheduling. Specifically, our proposed BiNoC architecture can improve traffic delivering efficiency and achieve the goal of power and area savings by increasing bandwidth utilization and reducing the physical volume of buffer memory. In summary, we provided a novel bidirectional channel NoC (BiNoC) backbone architecture, which can be easily integrated into most conventional NoC designs and successfully improve NoC performance with reasonable costs.

Appendix A
Simulation Environment

In this section, we will introduce our simulation platform, including hardware implementation environments and verification traffic patterns. All of the functional modules and performance results provided in this book were constructed and simulated based on the NoC platform as described in this appendix. Design modules described in different chapters of this book were broken down into respective components for rapid prototyping and customization.

A.1 NoC Platform

A cycle-accurate NoC simulation environment was implemented along with different inter-router channel designs in HDL. Each design comprises multiple functional blocks involving input buffer control, routing computation, virtual-channel allocation, channel control, switch allocation, and switch fabric. The physical layer of our simulation environment comprises 8×8 nodes connected as a mesh array. Design modules were broken down into respective components to allow rapid prototyping and customization. This environment can simulate on-chip interconnect networks using a packet-buffer and flit-buffer based flow-control protocol by dividing a packet into header, body, and tail flits. The number of virtual-channels and the depth of each input buffer can be easily configured for performance evaluation. Each packet had a constant 16-flits length. Also, basic round-robin principle was used to resolve the arbitration conflicts.

A.2 Synthetic Traffic

Four types of synthetic traffic patterns were used to run simulations, including uniform, transpose, regional, and hotspot traffics. In uniform traffic, a node receives a packet from any other node with equal probability. Every node transmits

S.-J. Chen et al., *Reconfigurable Networks-on-Chip*,
DOI: 10.1007/978-1-4419-9341-0,

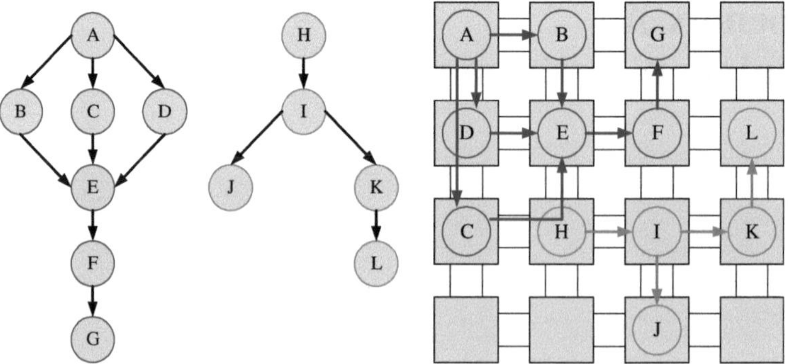

Fig. A.1 Task graphs of consumer benchmark and its mapping results

packets to evenly randomized destinations with a probability based on the injection rate. In transpose traffic, a node at (*i*, *j*) always sends packets to a single node at (*j*, *i*). In regional traffic, 90% of the packets are sent to the destination within three hops away. That is more realistic since each node does not often send packets to distant nodes. In hotspot traffic, 20% of the packets change their destination to selected hotspots with the remaining 80% of the traffic being uniform.

A.3 E3S Benchmark

The Embedded System Synthesis Benchmarks Suite (E3S) was designed for use in automated system-level allocation, assignment, and scheduling researches. There is one task set for each of the five application suites: automotive/industrial, consumer, networking, office automation, and telecommunications.

First, we input task graphs given in the E3S benchmarks and performed task mapping to map each task in the graph to a tile on an NoC. Task mapping will put tasks with heavy communication closer to optimize the communication cost. Then, the resulting process graph was converted to a packetized traffic flow which was fed into our NoC simulator. Here, we adopted the common *Simulated Annealing* algorithm as our task mapping algorithm. The cost function for mapping is the data flow between tasks:

$$\sum_{\forall w_{ij} \in task\_graph} w_{ij} \times \left(\left|x_i - x_j\right| + \left|y_i - y_j\right|\right),$$

where w_{ij} represents the total communication volume between $task_i$ and $task_j$ while (x_i, y_i) and (x_j, y_j) represent the positions of PEs which are assigned to $task_i$ and $task_j$, respectively.

Since the in-degree or out-degree of each task in real application such as EEMBC is not huge, most of the task pairs that have inter-task data communication can be mapped nearby. Take *consumer* benchmark in EEMBC for example, as illustrated in Fig. A.1, after an appropriate task mapping process, most of the connected tasks are mapped nearby and the traffic flow can be delivered using two bidirectional channels between their associated tiles.

Appendix B
Performance Metrics

Performance of a network can be generally described by a curve that depicts the relationship between average latency and flit injection rate (offered traffic) as shown in Fig. B.1. Traffic delivering latency stands for the time from when the first bit of the packet is moved into the interconnection network at its source terminal to when the last bit of the same packet arrives at the destination terminal. Offered traffic is the average amount of data traffic injected from each source terminal, which we usually measured in terms of average flit injection rate. This latency versus flit injection rate curve varies with different types of traffic patterns applied in the network.

In Fig. B.1, latency approaches the zero-load latency T_0 at an offered low traffic. Zero-load latency gives a latency lower bound under the condition that a packet needs not to contend for network resources with other packets. The zero-load latency T_0 can be divided into two terms.

$$T_0 = H_{ave}\, t_r + \frac{L}{b}$$

The first term is the head latency with an average hop count H_{ave} and a delay of single router t_r. The second term represents the serialization latency which is the time for one packet of length L to cross a channel with bandwidth b. Latency goes to infinity at the saturation throughput λs which is affected by network topology, routing algorithm, and flow-control of a router design.

Throughput is another performance metric which represents the accepted traffic rate at the destination terminals as illustrated in Fig. B.2. When a network approaches and exceeds its saturation throughput, the focus of the designer generally shifts from latency to the underlying fairness of the flow-control technique. As described in [1], if a saturated channel is not fairly allocated between flows, the network becomes unstable; that is, some flows will become starved and their throughput will drop dramatically as the flit injection rate increases beyond the saturation point.

S.-J. Chen et al., *Reconfigurable Networks-on-Chip*,
DOI: 10.1007/978-1-4419-9341-0,

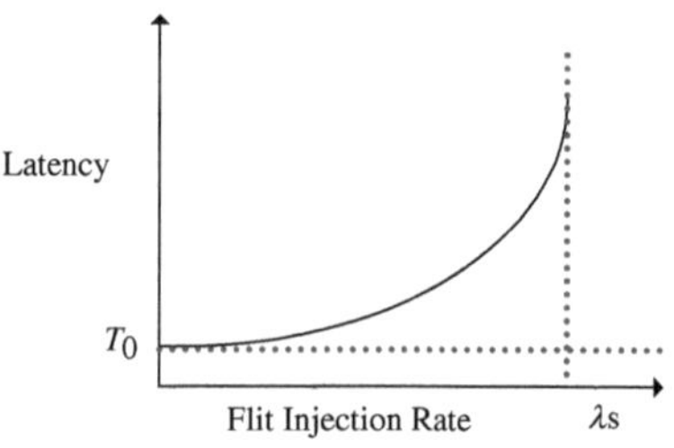

Fig. B.1 Latency versus flit injection rate for a network

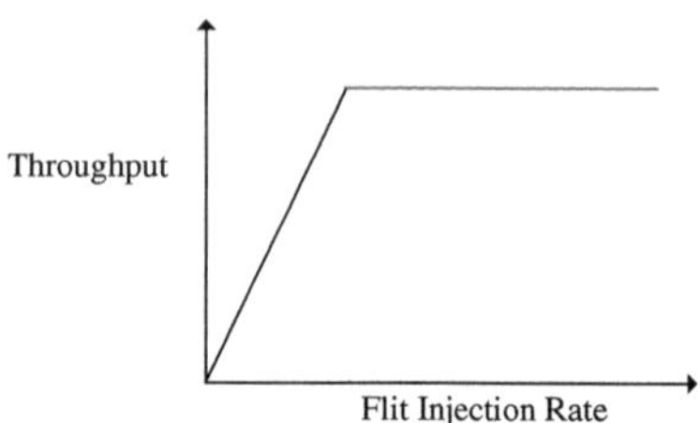

Fig. B.2 Throughput versus flit injection rate for a network

Reference

1. W. J. Dally and B. Towles, *Principles and Practices of Interconnection Networks*, Morgan Kaufmann, 2004.

Index

S.-J. Chen et al., *Reconfigurable Networks-on-Chip*,
DOI: 10.1007/978-1-4419-9341-0, © Springer Science+Business Media, LLC 2012

MIX
Papier aus verantwortungsvollen Quellen
Paper from responsible sources
FSC® C105338

If you have any concerns about our products, you can contact us on
ProductSafety@springernature.com

In case Publisher is established outside the EU, the EU authorized representative is:
Springer Nature Customer Service Center GmbH
Europaplatz 3, 69115 Heidelberg, Germany

Printed by Libri Plureos GmbH
in Hamburg, Germany